A Text Book Of

CONSTRUCTION MANAGEMENT

(22061)

Semester - VI

FOR

THIRD YEAR DIPLOMA COURSE IN CIVIL ENGINEERING GROUP

As Per MSBTE's 'I' Scheme Syllabus

VAIBHAO K. SONARKAR
B. E. (Civil), M.E. (Environmental Engg.)
Civil Engineering Deptt., ISTE (LM), IAPAC (LM)
V. B. V. Polytechnic, Vasai Road (W),
PALGHAR – 401202

Mrs. RUPALI P. KHADTAR
I/C Head, Department of Civil Engineering
Anjuman - I - Islam's Abdul Razzk
Kalsekar Polytechnic, PANVEL

DATTATRAYA J. KHAMKAR
M.E. Civil, M.P.M.
Department of Civil Engineering,
Trinity College of Engineering and Research,
PUNE

N4555

CONSTRUCTION MANAGEMENT ISBN 978-93-89825-05-3

First Edition : January 2020
© : Authors

Published By :

NIRALI PRAKASHAN

Abhyudaya Pragati, 1312, Shivaji Nagar,
Off J.M. Road, PUNE – 411005
Tel - (020) 25512336/37/39, Fax - (020) 25511379
Email : niralipune@pragationline.com

➢ DISTRIBUTION CENTRES

PUNE

Nirali Prakashan : 119, Budhwar Peth, Jogeshwari Mandir Lane, Pune 411002, Maharashtra
(For orders within Pune) Tel : (020) 2445 2044, Mobile : 9657703145
Email : niralilocal@pragationline.com

Nirali Prakashan : S. No. 28/27, Dhayari, Near Asian College Pune 411041
(For orders outside Pune) Tel : (020) 24690204 Fax : (020) 24690316; Mobile : 9657703143
Email : bookorder@pragationline.com

MUMBAI

Nirali Prakashan : 385, S.V.P. Road, Rasdhara Co-op. Hsg. Society Ltd.,
Girgaum, Mumbai 400004, Maharashtra; Mobile : 9320129587
Tel : (022) 2385 6339 / 2386 9976, Fax : (022) 2386 9976
Email : niralimumbai@pragationline.com

➢ DISTRIBUTION BRANCHES

JALGAON

Nirali Prakashan : 34, V. V. Golani Market, Navi Peth, Jalgaon 425001, Maharashtra,
Tel : (0257) 222 0395, Mob : 94234 91860; Email : niralijalgaon@pragationline.com

KOLHAPUR

Nirali Prakashan : New Mahadvar Road, Kedar Plaza, 1st Floor Opp. IDBI Bank, Kolhapur 416 012
Maharashtra. Mob : 9850046155; Email : niralikolhapur@pragationline.com

NAGPUR

Nirali Prakashan : Above Maratha Mandir, Shop No. 3, First Floor,
Rani Jhanshi Square, Sitabuldi, Nagpur 440012, Maharashtra
Tel : (0712) 254 7129; Email : niralinagpur@pragationline.com

DELHI

Nirali Prakashan : 4593/15, Basement, Agarwal Lane, Ansari Road, Daryaganj
Near Times of India Building, New Delhi 110002 Mob : 08505972553
Email : niralidelhi@pragationline.com

BENGALURU

Nirali Prakashan : Maitri Ground Floor, Jaya Apartments, No. 99, 6th Cross, 6th Main,
Malleswaram, Bengaluru 560003, Karnataka; Mob : 9449043034
Email: niralibangalore@pragationline.com

Other Branches : Hyderabad, Chennai

niralipune@pragationline.com | www.pragationline.com

Also find us on 🄵 www.facebook.com/niralibooks

Preface ...

This Text Book of **"Construction Management"** is strictly written as per the new syllabus of 'I' Scheme effective from June 18, prescribed by the Board of Technical Examinations, Mumbai, Government of Maharashtra.

The arrangement of the topics in the book is made in order to ensure a smooth flow of the subject for better understanding of the concepts. Along with theory, as subject is experiment based, details of experiments are also given in this book as per the need of the students. Solved examples have been included to get an idea of the numericals. So we hope that all need for this subject will be completely fulfilled in this book.

Our sincere thanks are to Shri. Dineshbhai Furia, Shri. Jignesh Furia, Shri. Pradeepbhai Furia and whole Staff of Nirali Prakashan, especially Mr. Shashikant Patel and Mr. Jayant Dedhia. We are also thankful to Mr. Santosh Bare (D.T.P.), Mrs. Manasi Pingle (Proof Reading), Mrs. Anjali Muley (Fig. Drawing) from Poona Office for making this effort possible.

Any suggestions and constructive criticisms for the improvement of this book are welcome.

Last but not least special thanks to our family members and well wishers for their patience and encouragement.

Authors

**Syllabus ...**

UNIT-I : CONSTRUCTION INDUSTRY AND MANAGEMENT

1.1 Organization - Objectives, Principles of organization, Types of organization; Government/Public and Private, Construction industry roles of various personnel in construction organization.

1.2 Agencies associated with construction work - Owner, Promoter, Builder, Designer, Architects.

1.3 Job layout for construction site.

UNIT-II : PLANNING AND SCHEDULING

2.1 Identifying board activities in construction and allotting time to it based on rate analysis, Methods of scheduling, Development of bar charts, Merits and Limitations of bar chart.

2.2 Elements of Network : Event, Activity, Dummy activities, Precautions in drawing network, Numbering the events.

2.3 CPM networks, Activity time estimate, Event times by forward pass and backward pass calculation, Start and Finish time of activity, Project duration, Floats, Types of floats - Free, Independent and Total floats, Critical activities and critical path.

2.4 Purpose of crashing a network. Normal time and Normal cost, Crash time and crash cost, Cost slope, Optimization of cost and duration.

2.5 Material management - Ordering cost, Inventory carrying cost, EOQ.

2.6 Store management, various records related to store management, Inventory control by ABC technique.

UNIT-III : SAFETY IN CONSTRUCTION

3.1 Safety in construction industry - Causes of accidents, Remedial and Preventive measures.

3.2 Labour laws related to construction industry.

●●●

Contents

•••

Chapter 1

CONSTRUCTION INDUSTRY AND MANAGEMENT

Syllabus

1.1   Organization - Objectives, Principles of organization, Types of organization; Government/Public and Private, Construction industry roles of various personnel in construction organization.

1.2   Agencies associated with construction work owner, Promoter, Builder, Designer, Architects.

1.3   Job layout for construction site.

Objectives

After learning this chapter, student will be able to :

- Identify the roles of different agencies in the given construction industry with justification.

- Prepare organizational chart for the given organization.

- Identify the functions of specified personnel in the given organization with justification.

- Prepare job layout for the given construction site.

1.1 ORGANIZATION

- **George R. Terry, a well-known management consultant defines an organization as follows :**

  o   Organization means 'a group of people working together to achieve a common goal'. A common goal in contractors organization is to complete the work as per tender documents as early as possible and to achieve maximum profit. A common goal in a factory may be to have maximum production.

  o   Organization is 'the arrangement of functions deemed necessary for attainment of the objectives and is an indication of the authority and the responsibility assigned to individuals charged with the execution, respective functions'.

  o   Organization is 'the process of identifying and grouping the work to be done'. It defines the delegates responsibility and authority. In an organization, people work together and effectively to reach the objectives. The work is distributed among the people, a practical shape is given to the objectives. Organization is subsidiary to management.

- There are three fundamentals of an organization. These are :

  (i)   People

  (ii) Work

  (iii) An organizational structure.

- The principle problem of an organization is to select the right type of people with required caliber to achieve the desired results. The problems of organization arises out of the division of labour.

  (i) People : People who carry out the activities of the organization must be well qualified and trained. They must perform the work given to them. They must be selected on the basis of their ability and skill.

(ii) Work : In the early days of industrial revolution, one man was doing all the work such as purchasing of goods, selling etc. Now, the work is divided into several departments e.g. purchasing, production, quality control, sales etc.

(iii) Organization Structure : There are four considerations :

(1) Span of supervision.

(2) Provision for making changes (flexibility).

(3) Balancing between the different departments.

(4) Integrating work with other phases of management.

1.1.1 Basic Activities in Organization

(1) Technical activities   --   Production, manufacture etc.

(2) Commercial activities   --   Buying, selling

(3) Financial activities   --   Optimum use of capital

(4) Security activities   --   Protection of property and persons

(5) Accounting activities   --   Stock takings, preparation of balance sheets etc.

(6) Administrative activities   --   Planning, organizing, controlling.

1.1.2 Essentials of Organization

- Organization is mechanism or structure that enables human beings to work effectively together. In the past, there were simple organsiations. But as life is becoming more varied and complex, organization of human beings is also becoming more and more complex.

- It is clear that, the utility and necessity of an organization consists in enabling the person concerned to attain the desired objectives and do their work efficiently.

- Whatever the type of organization, simple or complex, it must be based on the following three principles :

  (1) Division of labour

  (2) Source of authority

  (3) Relationships.

- Work should be properly divided between different persons and groups. Each person or group must be known what he or they have to do.

- There must be means to ensure that, the work assigned is done, and done properly. Each person in the group must be accountable to somebody and the leader of each group must also be made responsible to some authority.

- Relations between different persons in the same group or different groups must be clearly defined. Such a clear definition of relations between persons and groups will avoid bitterness, confusion and misunderstandings.

- An organization structure worth its name must satisfy the above three criteria. An organsiation lacking in one of these three principles will remain defective and may fail to attain the desired objectives.

1.1.3 Types of Organization

- The general organsiation and distribution of functions within a firm depend to a large degree upon the size of a firm and type of work generally undertaken by it.

- The following types or organizational structures are commonly met with :

  (i)   Military or line type organization.

  (ii) Line and staff organization.

  (iii) Functional organsiation.

(i) Military or Line Type Organization : This is the oldest and simplest form of organization. There is a leader and a few workmen. This is illustrated below :

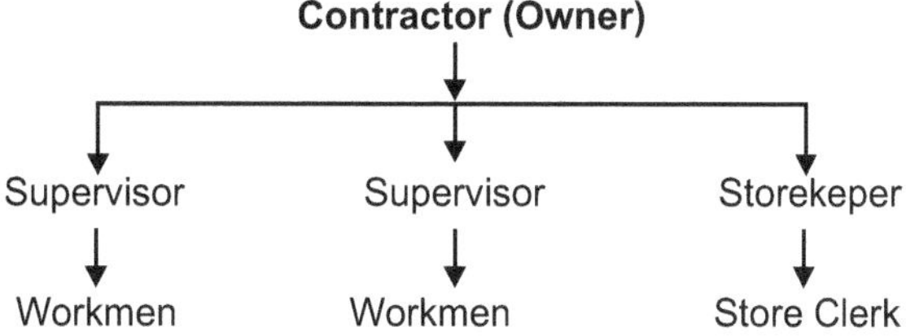

(a) Military or Line Organization :

- In this type of organization, the leader has the authority over the workmen and directs them in respect of what to do, when to do and how to do. The workers know where exactly they stand in relation to their superior. This helps to maintain discipline which is an essential and important factor in any type of organization.

(b) Expanded One Man Organization :

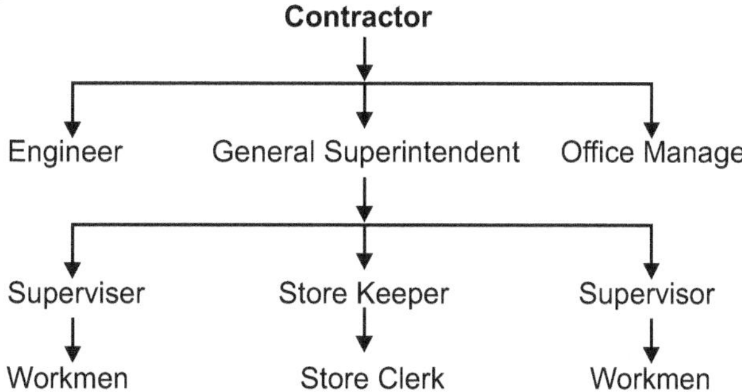

- In case of large and complex works one man alone is hardly able to perform varied duties, he is called upon to do and hence other people have to be brought into the organization. These additional persons have to be brought into an organization.

- These additional persons are specialists such as an Engineer, an Office Manager and General Superintendent. These men are acting as heads of their respective departments and the contractor acts primarily as an administrative head.

- The above chart shows a typical example of military or line type organization of a medium size firm engaged mainly in Civil Engineering and Public Works Construction.

Advantages :

(1) It is simple, easy to understand (easy to establish).

(2) There is clearcut division of authority and responsibility.

(3) Quick action is possible.

(4) Discipline is easily maintained.

Disadvantages :

(1) It is rigid and inflexible.

(2) There is lack of expert advise and services, because the organization lays too much emphasis on the ability and strength of a few able men.

(3) The division of work is done in accordance with the views of manager.

(4) Organization of this type does not have a scientific principle.

Applications :

This type of organization is applicable to the following organizations :

(1) Small organization having few subordinates.

(2) Routine type business organization.

(3) Mechanical organization.

(4) Labour management relations are co-ordial.

(5) Continuous production, simple and established organization.

(ii) Line and Staff Organization :

- The military or line organization was useful in the early stage of industrialization. But as the industries began to grow in size and complex began to be used in them, the manager found it difficult to manage the concern by himself. He was required to take the aid of men who possessed special knowledge and skill.

- In the line and staff organization specialists or experts are engaged but they serve in an advisory capacity and do not have any apparent authority. The experts have no authority to direct the workmen.

- Flow chart below shows a typical example of line and staff organization.

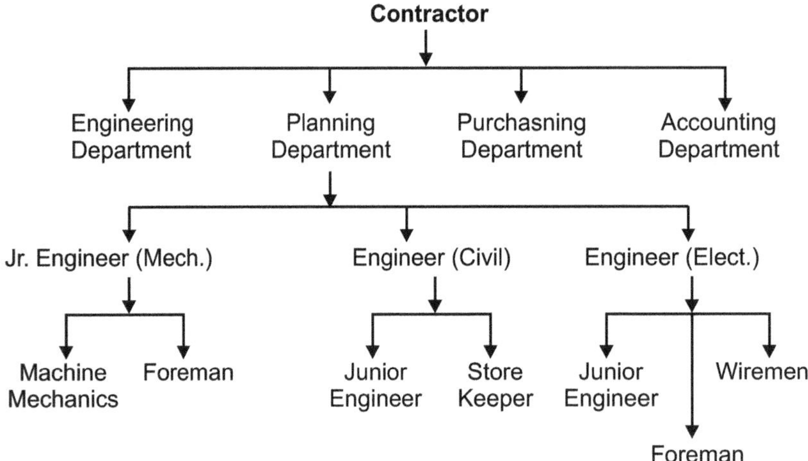

Advantages :

    (1) Expert advice is available at all levels.

    (2) The operations are efficient and hence economical.

    (3) The staff at site is competent to take decisions.

    (4) It also offers better facilities for expansion of business units.

Disadvantages :

    (1) There may be confusion about the relation of staff and line employees.

    (2) The staff may be ineffective for lack of authority to carry out its recommendations.

    (3) Line employees may resent the activities of staff members, feeling that the prestige and influence of linemen suffer from the presence of the specialists.

(iii) Functional Organization:

- In line and staff organization, the staff departments seldom restrict their activities to advisory or planning capacities. At times the staff officers may be more influential than some departmental heads.

- Thus, there is disparity between the theory and actual practice. This disparity is eliminated in functional organization wherein the authority of staff specialists over line officials at lower level is recognised. In all other aspects it is similar to the line, and staff organization.

Advantages :

    (1) Division of labour is done on functional specialisation.

    (2) Division of work is done according to the aptitude and hence increase in the sale.

    (3) Steady research and improvement of work is possible,

    (4) Standardisation of procedure and performance is possible,

    (5) There is a wide scope for expansion of business.

    (6) Economy in labour, time and money can be achieved.

    (7) Sense of responsibility is developed.

Disadvantages :

(1) It is difficult to define who the boss is.

(2) It de-emphasizes the position of the line organization.

(3) It may present special difficulties of co-ordination as :

    (a) Initiation on supervisors may become shifted and even routine work may become complicated.

    (b) Overlapping authority may give rise to friction between workers and the supervisors.

    (c) Too much of sub-division also leads to lack of co-ordination.

(4) Costly for management and used only in large organizations.

(5) There is no incentive and motivation for work.

(6) In case of unsatisfactory results, it is difficult to blame one foreman, and there are chances of avoiding responsibilities on the part of the foreman, various Civil Engineering agencies, their scope and functions users/promoters, Financers, Architects, Consulting Engineers, Designers and Builders.

1.1.4 Organization of Public Works Department (P.W.D)

- The various engineering departments of the government responsible for construction and maintenance of Public Works such as building, roads, canals, public health and electrical installations are known as Public Works Department.

- The main branches of P.W.D. are :

(1) P.W.D. (Public Works and Housing Department).

(2) P.W.D. Irrigation.

(3) Environmental Engineering Department.

- Public Works Department may be central P.W.D. (under the Government of India) or State P.W.D. (under State Government). For example, a building for a post-office will be built by the Central P.W.D., but a bungalow of a Principal, Government Polytechnic in Maharashtra State will be built by the State P.W.D.

- The other departments of government are :

(a) The Railway Department.

(b) Military Engineering Services - Defence Works.

- Though the above two departments do not strictly fall in the category of P.W.D. the working of these departments is very much similar to the working of other P.W.D.'s.

- The other agencies which deal with construction of roads, buildings etc. fall in the third category, i.e., Semi-Government department such as :

(1) Municipal Corporations e.g. Mumbai, Pune.

(2) Maharashtra State Electricity Board.

(3) M.I.D.C.

The working of these Semi-Government departments is similar to that of P.W.D.

1.1.5 Administrative Set-Up of Public Works Department

The following chart gives the normal set-up of the department of the Centre or the State as the case may be :

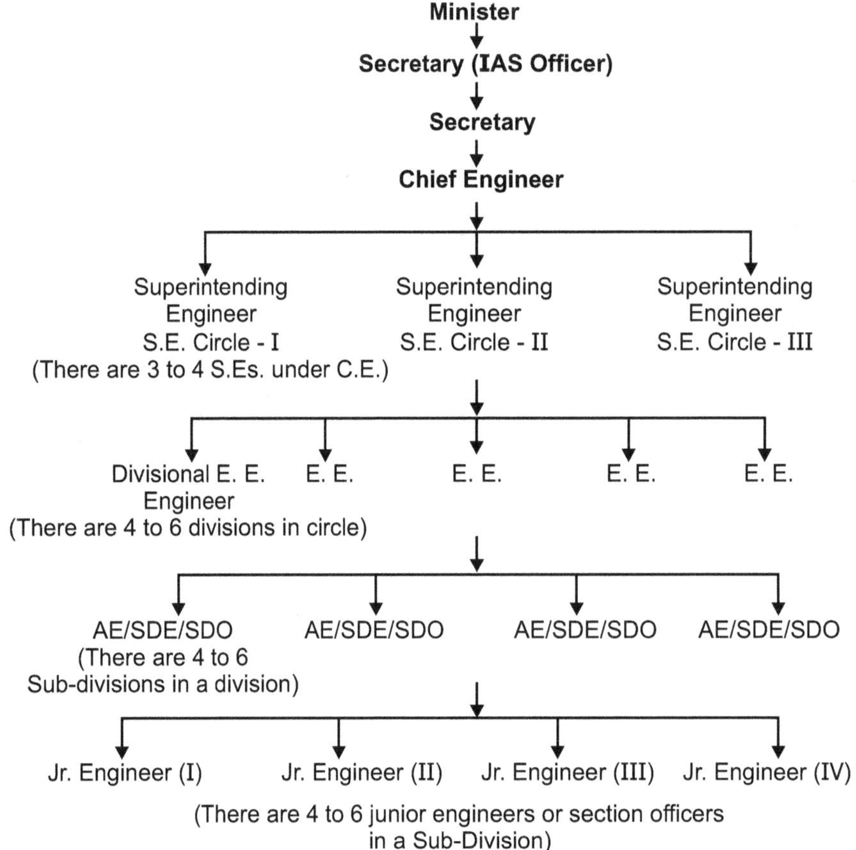

Functions of the Chief Engineer :

- Chief engineer is the technical head of the department. He submits the budget proposals to the government in respect of all original and maintenance marks in his jurisdiction and exercises control over the expenditure in collaboration with Accountant General of the State.

- As a head of the organization he enjoys the following financial and administrative powers/authority.

  (a) Financial :

  (1) Administrative approval of works other than buildings upto ₹ 25,000/- (This may increase from time to time).

  (2) Acceptance of tenders for works by contract - full powers.

  (3) Sanction Deposit works - full powers.

  (4) Purchase of live stock ₹ 30,000/- at a time.

  (b) Technical :

  To accord technical sanction for original work and repairs upto 5 % in excess of the administratively approved amount.

Functions of the Superintending Engineer :

- Superintending engineer is responsible to the Chief Engineer for the following in respect of his circle :

  (a) Administrative and Technical control of the circle.

  (b) Exercises financial control over execution of original and repair works.

  (c) Inspects works in his circle.

(d) Arranges procurement of stores.

(e) Inspects divisions in his circle and reports to C.E.

(f) Exercises administrative control in respect of Technical staff.

(g) Keeps himself informed about the progress of works in his circle.

Functions of the Executive Engineer :

- The Executive Engineer is responsible to the Superintending Engineer, for administrative and technical control of his division.

His main duties are :

(1) To see that no work is taken in hand without the administrative approval and allotment of funds and technical sanction.

(2) To exercise effective supervision on the works in his division.

(3) To see that contractors' accounts are properly maintained and are regularly submitted to the competent authority.

(4) As a drawing and disbursing officer of the division he draws money from the concerned treasury and makes payment to the contractor and labour and office establishment against funds placed by the government in the concerned treasuries.

(5) Carries inspection of the offices of AE/SD /SDO atleast once in a year.

(6) Ensures that all tools, plant and machinery are properly maintained.

Functions of AE / SDE / SDO :

- Assistant Executive/Engineer, Assistant Engineer or Sub-divisional officer is responsible to the Executive Engineer for efficient management and execution of works in his sub-division.

Duties :

(1) Arranges stock, execution, progress and completion of all original works and repairs in the sub-division.

(2) Checks works in progress.

(3) Maintains construction accounts.

(4) Physical verification of measurement books.

(5) Takes proper care of cash and stores.

(6) Physical verification of dead stock once in a year.

(7) Checks muster rolls.

(8) Maintains all buildings/roads.

(9) Exercises administrative control of the office and field staff in the sub-division.

Functions of the Junior Engineer :

(1) Supervision of all works in his charge executed through contractor or departmentally.

(2) Maintains attendance of departmental works.

(3) Maintains attendance of departmentally employed daily labour on muster roll, and of casual labour, either directly or through the field staff consisting of sub-overseers.

(4) Takes measurements of all works in the measurement books.

(5) Exercises direct control in respect of specifications and progress of works under his charge.

(6) Prepares running bills and final bills for all works in his charge.

(7) Prepares pay-bills of the work charged establishment of his section.

(8) Carries out field survey of proposed projects and ground tracing of structures on the ground.

(9) Prepares estimates for all types of works in the section.

(10) Carries out check of tools, plant etc., arranges collection of rents and revenue due to the government,

(11) Carries out half yearly checks of all stores in his charge and submits reports to S.D.O.

(12) Carries out annual inspection of all structures in his charge and submits reports of such inspection to the S.D.O.

(13) Initials disposal of unserviceable stock, tools and plants in the form of survey report.

1.1.6 Organizational Structure of a Company

- A company's organizational structure is by and large determined by the objectives of the company and the work involved in reaching them. Therefore, the organizational structures varies from company to company.

- Marketing and Finance are common functions to most of the business organizations and production is common to all manufacturing companies.

- Division of work is one important phase of setting up an organization. But it is very necessary to provide for good co-ordination in large companies, special co-ordination committees. General staffmen can render useful work in this regard.

Delegation of Authority :

- A key problem concerning an organization is the degree to which authority should be held or dispersed throughout the organization. The main question is how much authority can be delegated to lower ranks.

- To the extent that authority is not dispersed, it is centralized. There cannot be an absolute centralization of authority.

- In that case, the status of a manager is lost.

- Centralization and decentralization are tendencies or qualities.

Advantages of Decentralization :

According to the findings of Peter Drucker, the following are advantages of decentralization.

(1) Speed in decision making and absence of confusion.

(2) No conflict between the top management and other ranks.

(3) Absence of politics, absence of fearness will prevail.

(4) Informal way of thinking, and democratization in the organization.

(5) Availability of potential managerial manpower.

(6) Absence of a gap between the few top managers and many subordinate officers in the organization.

Decentralization is indicated in the following ways :

(1) To lower burden on top executives,

(2) To facilitate diversification,

(3) To provide product and market emphasis,

(4) To encourage development of managers and

(5) To improve motivation.

- As mentioned earlier, there cannot be absolute centralization or absolute decentralization. Decentralization can be effective only if it is balanced by a suitable degree of centralization.

- There must be necessary centralization of planning, organization, motivation, co-ordination and control.

1.1.7 Development of an Organization

- In the development of an organization, the following points are to be considered :

(1) The objectives of the enterprise or company must be well defined and determined.

(2) The areas of activities must be created.

(3) A cohesive organizational structure must be created.

(4) A survey of personnel available must be undertaken and services secured.

(5) The structure must fit into the available personnel and other resources. The organization must be flexible, it has to accommodate changes, when required. But, this does not mean haphazard changes.

- Leadership is very important for an organization. It is something more than personal ability and skill. A true and good leader must be intelligent, competent, responsible and dynamic. A weak organization with a strong leader may survive and run better than a strong organization with ineffective leadership.

1.1.8 Organization Suitable for Construction Firm

- The organization for a construction firm has to be developed taking into account the general principles of organization and the special characteristics of the construction industry.

- It is difficult to laydown specific rules for organizing the structure of a construction firm. It depends on the nature of activities, scope and type of project work and the areas of operation of the company or firm.

- The type of organization suitable for a small construction firm is a simple line or line and staff type of organization as shown in the chart below.

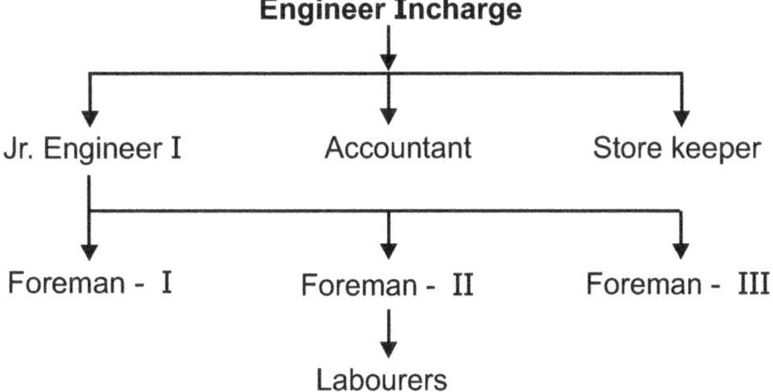

- The engineer-in-charge at the site exercises full authority and is responsible for the execution and progress of work. Under him are a few section officers or foremen who get the work executed. In a very small work such as construction of a private residential house, the contractor has to himself perform the functions of engineer.

1.1.9 Types of Contracting Firms

- The following are important contracting firms :

  (i) Sole proprietorship,

  (ii) Partnership,

  (iii) Private limited company,

  (iv) Co-operative societies :

1. Sole Proprietorship :

- In this type of contracting firm, the individual proprietor organizes and manages the business himself. He is fully authorized and responsible for every thing including loss or profits.

- This type of firm is suitable when the business is small and limited which can be managed by an individual. It does not involve much legal formalities or other complicated procedure.

  Advantages :

  (a) Simple and easy.

  (b) Least legal formalities.

  (c) Quick decisions.

  (d) Quality production.

(e) Better labour relationship.

(f) Personal attention.

(g) Small capital.

(h) Maintenance of secrecy.

Disadvantages :

(a) Limited capital.

(b) Personal limitation.

(c) Cannot compute with large contracts.

(d) Short life.

2. Partnership firm :

- When the business increases, it becomes difficult for an individual to manage. It then becomes essential to take some one, or more partners. Partnership business is owned by two or more persons (up to 20) who share the powers, responsibilities and profits according to an agreement made among themselves i.e. Partnership deed.

  Advantages :

  (a) More capital.

  (b) Diverse Talent.

  (c) Less possibility of errors of judgement.

  (d) Prompt decision.

  (e) Large economics.

  (f) Personal factors.

  (g) Division of labour.

  Disadvantages :

  (a) Unlimited liability.

  (b) Dis-agreement.

  (c) Less secrecy.

  (d) Non–transfer of partnership.

3. Private limited company :

- This type of firm can be formed by two or more members. The maximum number of members is limited to 50.

- The company is registered under the Indian company Act 1956. The transfer of shares is limited to members only and general public cannot be invited to purchase the shares. Most of the middle size industries are run in this manner.

4. Public limited company :

- As its name indicates, the membership of public limited company is open to general public. The minimum number of parsons required to form a public limited company is seven, but there is no upper limit.

- Such companies can advertise to offer its shares to general public. Such companies are subjected to greater control and supervision of the Government. This control is necessary to protect the interest of the share holders and the members of the public.

- Shares of such companies are transferable. The affairs of the company are managed.

5. Co-operative organsations (Societies) :

* **Definition :** A co-operative society is a voluntary association of economically weak persons who work for achivement of their common economic objectives on the basis of equality and mutual service.

* The main object of co-operative society is to promote self help and mutual assistance among men of moderate means and income having needs and interest in common. Such men are industrial workers, small artisans, agriculturist and members of the middle class.

* The members supply the capital, manage the business and share all profits and losses.

Formation of Co-operative Society :

* To start a co-operative society, an application is submitted to the registrar of co-operative societies. The official of this co-operative department will attend the first general body meeting in which bye laws are formed to govern the society and the directors are elected by the share holders.

* If the authorities are satisfied about the soundness, a licence will be issued by the registrar and the co-operative society is formed.

* The various types of co-operative societies are :

    1. Producers co-operative society.

    2. Consumers co-operative society.

    3. Housing co-operative society.

    4. Credit co-operative society.

Advantages :

    (a) Provides better methods of execution of small contracts.

    (b) Elimination of middle man.

    (c) Democrative nature.

    (d) Sense of co-operation.

Disadvantages :

    (a) Lack of co-ordination.

    (b) Chances of undue advantages.

    (c) Favourism.

    (d) Limited capital.

    (e) Political influence.

1.2 PRINCIPLES OF ORGANIZATION

* The following are the principles for some organizations. Many principles have been enumerated by management scholars such as Taylor, Fayol, Burnord. Many of these principles have been applied by well managed American companies.

    (1) Part of responsibility and authority : One cannot exist without the other. A person cannot be held responsible for duties if he does not have the power to do. Responsibility cannot be delegated or shifted. The responsibility of a superior for the action of his subordinates is absolute.

    (2) Balance : In every organization there must be balance. For example, balance between centralization and decentralization.

    (3) Flexibility : An organization should not resist to changes those are necessary due to technological changes, economic changes, personal changes etc.

    (4) Continuity : The organization must provide means for effective continuity.

    (5) Leadership : The organization must create a situation where managers can effectively lead. Organization is technique of leadership in the above sense.

Division of Work :

- This has been mentioned while studying Fayol's principles of management. This is the principle of specialisation which is to produce more and better work with the same effort.

Functional Definition :

- Departmentation is the grouping of activities to make the organization effective and efficient. The activities must be defined. There must be clear specification of duties and of authority relationship.

  Unity of Command : This is already discussed, there must be only one immediate boss or superior. A worker should not receive instructions from several individuals.

  Unity of direction : There must be one head and one plan for a group of activities having the same objective.

  Exception of principle : Each manager at each level has to make decisions in the light of his authority. If he cannot decide because of limitations of his authority, must refer to his superiors.

  Span of control : This denotes the number of individuals, one supervisor (or manager) directs. There is a practical limit to this. A large number is not good as effective supervision is not possible. A small number means the supervision time is not effectively used. An appropriate number must be found.

  Communication : This is vital for an organization to succeed. Generally, the lines of authority serve to provide channels of communication downwards. Upward communication is blocked or not encouraged. These must be encouraged. There must be two way communication (top boss to the lowest rang and from workers to the top bosses.)

Contributions Made by the Civil Engineering Profession :

- Civil Engineering is one of the oldest professions ever practised by the civilized world. Historical evidences show that the earliest structure built were the civil engineering works.

- During the last few centuries, vast strides have been made by people all over the world in all spheres of human activity. Man has explored the frontiers of nature and has harnessed the natural elements.

- The achievements of man in the field of engineering are phenomenal. Innumerable innovations are made almost everyday. The socio-economic conditions are rapidly changing the structure of our society.

- The advances made in the field of civil engineering are too many and very striking. The profession has offered the civil engineer great challenges to his genius. He finds his rewards in creative work, in the service of his fellowmen and in the making up of a better world for all to live in.

- The economical growth of a country is related to the growth of engineering and technology. Civil Engineers can take pride for their lofty contributions made to the economic progress of the world.

- We can cite the example of our own country. If we look into the investments made in the various five year plans, we find that Civil Engineering works have taken a lion's share.

- Even for steel plants, thermal station, atomic power station, we require civil engineering works.

- In fact, it is difficult to think of any industrial or economic activity without civil works figuring in there. Thus, civil engineering works are the basic needs of our society. Whether it is war time or peace time, these works go on for ever.

- There can not be a bull in the activity of a civil engineer. If it happens it will create economic stagnation and material prosperity will be halted. All the activities of the civilized world will be adversely affected. The economic order of our society is linked with the civil engineering works.

Range of Civil Engineering Field :

- The Civil Engineering world covers a wide range, a very vast field or domain. It includes diversified fields of activity such as irrigation, public health works, construction on bridges, docks harbours, airports, tunnels, highways and railways, hydro-electric work, nuclear power stations and building activities of a kinds.

- Each of these activities is concerned with the welfare of our society and growth of economic prosperity. New methods and techniques are used to make improvements in the organization, planning and performance of construction work.

- Research and development works are taken in an ever increasing way to evolve new methods, cheaper methods and safer ways of doing things. This sound management principles are being applied to the construction industry.

- In present days civil engineer (or the manager) makes use of the management tools or techniques. Because of the complex nature of construction projects, it has become all the more necessary to apply management techniques to achieve increased productivity, efficiency and economy.

Nature of Civil Engineering Works :

- Civil Engineering works, although organized on an industry basis, are different from those of mechanical, electrical and other fields. Civil Engineering works, by and large, are executed in the field. The works can be compared to factory made products.

- Every civil engineering work is unique in its own way. No two works are identical. Every new work is built in a new location with different environments, with different site conditions. While factory made products are of a repeatative nature, civil works are far from them.

- Whether it is the construction of a bridge or tunnel or a building, we cannot obtain the same site conditions every where. Each work and each site present new problems. We cannot have a tailor made solutions for every construction.

- Every site presents a new situation, new conditions, new problems. Each requires a different or individual tackling, distinct approach, a new method. Even working methods cannot be alike, equipment cannot be same. Such is the unique nature of Civil constructions.

Nature of Labour Force :

- One of the most distinguishing feature is the ever changing labour force. More often and then, not a new labour force is employed for every new work taken up. On completion of a project or work, the labour shifts to a new place or site in search of work.

- It rarely happens that an owner employs the same people. With individual owners, this does not arise at all. On completion of their building or house construction they do not need the men employed by them. The workers go in search of a new employees. This is true of all private works.

- Sometimes, some contractors have permanent gang of workers, working for them. It is likely that, most of them may work for the same employer. But, these cases are not many.

- If the works are situated at far off places or in different towns and cities, the labour may be reluctant to shift or move from place to place. Even in the case of big Government works (such as irrigation projects) the government employs a very large number at the peaks of construction activity.

- Unless similar big works are taken up immediately, there is no use for this large work-force. Government cannot employ them. This is a very important point to note.

Comparison with Factory Works :

- Compare the situation as it exists in a factory. The work force is for all practical purposes, of a permanent nature. The management may take up new orders, new work every time.

- But, the same workers continue to work, under the same shed, same premises, day after day, year after year, perhaps tending the same machines or equipment.

- Workers do not leave the work place unless they want to change their job or they are on work-charged establishment, or daily wages and want to change as management is likely to discharge them.

- With this situation obtained in a factory or business establishment, it is possible for the management to take up various programmes such as training, conducting of special courses for the workers.

- In other words, they can create opportunities for the workers to improve their skill, efficiency and knowledge. There is scope for improvement, for development and for promotion to higher ranks.
- With ever moving nature of civil engineering work force, none of the above programmes is possible. A civil engineer deals every time with a new task force.
- If the work is new or workers are new to the job, the site engineer has to exert, train his men before they can be effectively employed. In Industry, due to labour turnovers there should exist, training before they can be effectively put on job. New faces come for the job and the foreman or supervisor has to exert. But such situations are not frequent.

1.2.1 Personnel Management

- For the efficient use of all other resources, efficient use of human resources in employment is essential otherwise raw materials will remain stocked, machines will remain idle and money will be blocked up and work will come to a standstill. Hence, management of man power has greater importance.
- **Definition :** Personnel management is that part of management which is specifically concerned with the people in the organization.
- It has to establish and maintain sound personal relations at all levels in the organization. It has to secure the effective use of personnel of various categories by ensuring such employment conditions which will enable all the people working in the organization to contribute most effectively to the objectives and also to achieve their personal and social satisfaction side by side.
- 'Men' is the most important factor of production but very complicated part of the duties of the manager. Men differ from material and machine in many ways as one should not forget that, they are human beings.
- Moreover, all individuals are different from one another. They react differently, think differently and behave differently in similar situation. Human tendency can not be precisely gauged.
- Every organization must possess good and efficient persons on its roll. No organization can ever achieve its aims and objectives in the absence of hard working, devoted, sincere and dynamic persons.

1.2.1.1 Functions of Personnel Management

The functions of Personnel Management are listed below :

(a) Procurement and Maintenance of adequate work-force (employees) as regards to both, the number and quality of personnel.

It involves :　　Recruitment,

　　　　　　　　Interviewing,

　　　　　　　　Testing,

　　　　　　　　Induction,

　　　　　　　　Placement,

　　　　　　　　Follow up of new employees for adjustment,

　　　　　　　　Merit rating,

　　　　　　　　Promotions, transfer and discharge,

　　　　　　　　Employment records.

(b) Education and Training of present employees

It involves :　　Job instruction,

　　　　　　　　Apprentice training,

　　　　　　　　Economic education,

　　　　　　　　Reading rooms and libraries,

　　　　　　　　Records of statistics,

　　　　　　　　Training by audio-visual aids.

(c) Maintaining satisfactory personal contacts and relationship :

It invloves : Job analysis and job specification,

Wages and rewards,

Labour record and statistics,

Regularisation of employment,

Labour turn over,

Morale studies.

(d) Maintaining satisfactory group relationships, by :

Contacting employee's groups,

Contacting employees representatives,

Contacting government agencies,

Integrating group interest.

(e) Maintaining employees health

It involves : Healths standards,

Sanitation control,

Treatment of minor injuries and diseases,

Rest period, recreation etc.

(f) Maintaining employee's safety

It involves : Safety guards and inspection of safety equipment,

Safety programmes, safety records,

Fire protection.

(h) Maintaining employees service welfare.

It involves : Savings and investment plans,

Group insurance,

Pension,

Housing programmes,

Recreation plans.

- Personnel management mostly deals with the selection and placement of the right men on the right jobs at the right time, providing them 'job satisfaction'. Keeping them motivated for all the time. It is very essential to adopt clear personnel policy and proper procedure of recruitment, training, rigid rules and regulations of service conditions.

- Personnel manager is mainly concerned with the following :

(1) Manpower planning;

(2) Recruitment;

(3) Placement;

(4) Training;

(5) Motivating;

(6) Performance appraisal;

(7) Wage standardisation and rationalisation;

(8) Public relation;

(9) Welfare.

1.2.1.2 Personnel Required on Construction Work

- For execution of work following persons are required at site :
  (1) Site Engineer
  (2) Supervisors
  (3) Mistries
  (4) Operators
  (5) Labour.

1. Site Engineer : He is the overall incharge of the execution of work at site. His decision is final in the execution of work. He has to interpret the designs and drawings and issue instructions accordingly. He has to employ the required staff. He has to see the procurement of adore and to decide the use of equipment and machinery. He is fully responsible for the satisfactory completion of the work. He has to co-ordinate the various agencies required for the work.

2. Supervisor : Junior engineers, Overseers, Mistries are the persons appointed for the supervision of the work. Their specific job requirements are :

  (i)   To supervise the work carried out by the skilled and unskilled labourers and to guide them whenever necessary, in technical matters.
  (ii) To prepare the schedule of work : Daily-work programmes under the direction of the Site Engineer and to arrange accordingly the men, materials, equipment for the execution of work.
  (iii) To check and see that, work is being carried out as per the drawing and specifications.
  (iv) To take the measurements of work done, record it, and prepare the interim payment bill.
  (v)   To keep the top officers informed from time to time regarding the progress of work, requirement of materials and difficulties in execution, if any.
  (vi) To keep a constant watch on materials required, stock available and see that, it is being properly utilised without wastage.
  (vii) To pass the instruction, to lower level, which are received from the top.
  (viii) To ensure safety and security at the work site.

1.2.2 Labour in Construction Industry

- Construction industry is one of the largest industry in India. More than five crores of workers are working in this industry. Industry does not possess permanent place of work, workshop and no permanent labour.
- The workers are hired as and when required (at times temporarily) and retrenched when their need ceases. The output of such labour is very low compared with that of labour in construction industry in foreign countries. This may be due to low emoluments, illiteracy and poor working conditions in our construction industry (in India) in particular.
- This type of labour could not organize and therefore, has no or practically very poor collective bargaining power and that is why workers are paid low wages. Besides low wages they are forced to live in crowded and unhealthy conditions in huts, among unhygienic surroundings with hardly any basic amenities of life.
- To overcome such a state of affair, the Central and State Government have brought out laws to regulate and improve the working conditions of labour. Some employers in the construction industry have also provided a number of welfare benefits to their labour, these days.

Type of Labour :

- The following types of labour are generally found working in construction industry :
  (1) Unskilled labour.
  (2) Semi-skilled labour.
  (3) Skilled labour.

1.   Unskilled labour : It is a type of labour who can carry out simple manual work such as cleaning, sweeping, digging, concrete mixing, ramming, curing, material handling etc.

Examples : Male mazdoor, Female mazdoor.

2.   Semi-skilled labour : It is a type of labour who can carry out operations which do not require high skill and training and who can operate conventional type of machinery. Such labour is produced from an unskilled labour who has been associated with a specific type of work over long period. No formal training is given to such labour. They possess some skill by their own experience of work.

Example : Concretor, Glazier, Bhistie, Scoffolder, Equipment and Machine attendant etc.

3.   Skilled labour : It is type of labour who can carry out works which need special skill. Such labour has to take specialised training. Such labour is produced after training programme in training institutes. Some persons acquire the skill by family traditions or by self-experience.

Examples : Mason, Carpenter, Brick-layer, Painter, Fitter, Bar-bender, Tiler, Plumber, Welder, Electrician, Truck-driver, Operator, Mechanic etc.

1.2.3 Work Done by Department and Work Given on Contract

Work done by department	Work given on contract
1. Work is carried out departmentally by employing daily labour as masons, coolies, bhisties, carpenters etc.	1. Work is got done through contractors who employ workers required for the completion of the work.
2. Materials required for construction tools and plants required for the operations are issued from store by indent or purchased.	2. Material, plant and equipment are to be arranged by contractor.
3. Payment is done by department directly.	3. Payment is done to contractor.
4. Quality checking is done by department directly.	4. Contractor is responsible for quality of work.
5. Work is not speedy.	5. Work is speedy.

1.3 ROLE OF CONSTRUCTION INDUSTRY IN NATIONAL DEVELOPMENT

- India has attained her political freedom, however, her economical freedom is yet to be achieved. Statements are formulating plans and projects - mainly Civil Engineering projects such as dams, canals, roads, railways, bridges, buildings etc., for the development of the country.

- But these projects should be completed effectively, satisfactorily and economically by Civil Engineering industry to help to achieve economical freedom.

- We have already gone through a series of five year plans. India has already made outstanding improvements in the field of industry development. However, to improve the standard of living of the common man and to be self-sufficient in most of our requirements - civil engineering projects such as Irrigation, Hydro-power, Road, Railway project have to be provided to the maximum possible extent in the next plans.

- There are number of burning problems of our country, to illustrate a few - shortage of food, fuel, unemployment, facilities for agricultural activities, cottage industries, drinking water supply in rural areas and so on. All these problems necessitates provision of more and more civil engineering projects in future.

- In developing countries, construction activities play a very important role and consume major share of the budget. It will be seen from our record that out of the total six - five year plans which have been completed so far, more than 50% of the finance is spent only on construction projects.

- The construction activity is bound to increase in near future. People are demanding irrigation projects, new road projects, new railway lines and residential and industrial buildings. There is going to remain tremendous housing shortage in our country. About 40% people reside on foot-path and in huts in a very bad and undesired conditions.

- Considering all these requirements of the people and the nation, large number of construction projects are required to be provided. This needs very large investment of finance, man power and other resources.

- The construction of such projects need for managing the manpower, procurement and flow of materials, finance, tools and equipments and sequencing the activities, co-ordinating the various agencies in such a way as to complete the work in optimum time and at optimum cost.

- In executing the projects, the engineers have to handle various types of workers. Worker's satisfaction at the job determines his productivity. Its emphasis is on the fact that emotions control worker's moral and efficiency more than money and amenities and therefore the prime important job of an engineer is to achieve desired ends by means of collective efforts of a group of human beings.

- There are many other fact of management but this fundamental requirement to organise human being is its distinguishing feature.

- During the last 30-40 years several innovations leading to many changes in the design, construction and maintenance have taken place. Consequently new materials, machinery, tools and equipments have also been brought into use. All these changes will have to be taken into account for economic construction and better results.

- New tools of management such as bar-charts, critical path method and programme evolution review technique are available for planning and monitoring construction activities.

- All the above factors necessitate a sound knowledge of construction management to plan, co-ordinate and organise the various activities for effectively and satisfactorily completion of the work.

- The huge magnitude of work involved particularly in major project has necessitated the study of construction management, application of modern principles, techniques, methods, operations and tools of construction management.

1.4 DEVELOPMENT OF SMALL SCALE CIVIL ENGINEERING INDUSTRIES IN INDIA

- Now, there is a remarkable change in the policy of the Government towards Small Scale Industries. It is observed that in view of the peculiar conditions, especially of population, availability of resources etc., in our country, small scale industries have a very vital and central place in the rapid development of industries.

- For the achievement of economic freedom government has, therefore, made in its industrial policy statement a clear announcement of the importance of such industries and has also given ample proof of it in practice in the development programmes which have been included in the Five Year Plans.

- Small Scale Industries are small units. They use power and small machines and employ a small number of workers.

- According to the Government of India's latest classification, small scale industries include units whose investment in plant and machinery does not exceed ₹ 10/- lakhs. They also include those small ancillary units whose investment in plant and machinery does not exceed ₹ 15/- lakhs.

- Such Small Scale Industries are very common in foreign countries like Japan, Switzerland, Sweeden. They exist side by side with the large manufacturing industries. In Japan 58% of the industrial population still gains its livelihood in small scale industries. In U.S.A., it has been estimated that small business makes up 93% of U.S. business establishments, employs 46% of the country's workers and handles 35% of volume of business.

1.4.1 Advantages of Small Scale Industries

1. Employment Potential : Small Industry provides more employment and this is a very important factor for our country where millions of people are either unemployed or under-employed.

2. Capital - Light : Small Scale Industry needs relatively smaller amount of capital than that required by large scale industry. As capital is very scarce in an under-developed country like India, Small Scale Industry requiring less capital occupies an important place.

3. Self-Employment : Small Scale Industry provides limitless opportunities for self-employment.

4. Quick Yielding : Small Scale Industries start yielding within comparatively short time.

5. Capital Formation : A person, starting his business will collect capital from his own source, from relatives and friends. This capital probably would never have come into existence.

6. Decentralization : The rise in Small Scale Industries will bring decentralization of industries and will thus promote the object of regional development.

7. Relieving Congestion in Urban Areas : By providing gainful employment in the rural areas, flow of population to Urban areas can be controlled and congestion in Urban areas can be changed.

1.4.2 Role of Small Scale Industry

Small Scale Industries play very important and crucial roles in a developing country like India to solve many burning problems such as :

1. To reduce the unemployment.
2. To increase the national income.
3. To reduce the shortage of consumer goods.
4. To accelerate the economic growth.
5. To promote regional development.
6. To reduce the inequalisation in the distribution of income and wealth.
7. To reduce over-crowding in urban a areas etc.

1.4.3 Small Scale Construction Industries

1. Extraction of engineering material from nature such as soil, murum, sand, limestone etc.
2. Manufacture of engineering materials such as bricks, tiles, pipes etc.
3. Casting cement-concrete products like beams, slabs, column, grill work monate blocks etc.
4. Wood-work, doors, windows, cupboards, frames etc.
5. Steel work, steel grill works, door, window-frames.
6. Supply of engineering materials.
7. Housing as an industry.

- In India, Small Scale Industries are in small number and in very small size. The number is required to be enhanced so as to have giant contribution to the national economy. Especially in Civil Engineering works, very few small scale industries are registered and very few have been making a remarkable contribution to export.

1.5 NATURE OF CIVIL ENGINEERING MANAGEMENT

- Civil Engineering is defined as "the art of directing the great sources of power in nature for the use of and convenience of man". For many centuries, the art of Civil Engineering is used by a man for design and construction of dams, roads, bridges, buildings etc.
- Civil Engineers are applying their skills to raise the general standards of living of the human beings.

1.5.1 Special Characteristics of Civil Engineering Work

1. Each structure is unique :

- Every Civil Engineering structure is unique. It has got its own specific condition. Foundation characteristics, design considerations, construction aspects, the purpose of two similar structures may also be different.

- Consider, driving the piles of identical dimension and same material. The purpose is usually same, namely to transmit the load through soft ground to hard ground.

- But there is a wide variety of soft grounds and it is most unlikely that any two sites will contain identical strata. There may be variation in access and working space and it is found that each site has its unique phenomenon.

- Even when similar houses are built in one area, foundation condition may differ, ground levels may be different and hence each structure is different from others, each is unique. Hence, assembly line technique cannot be adopted here.

- Normally, there is a great deal of difference from building to building, bridge to bridge and generally from one construction to another construction. So every new construction is unique and has to be tackled separately. Because of these characteristics, the Civil Engineering construction management has to be more flexible and innovative.

2. Each structure is commissioned :

- For every structure of Civil Engineering an order is placed before construction commences. Civil Engineering firms, never initiate construction projects. To construct a new work, a decision is taken by others. The industry thus cannot be master of its own destiny.

- In other industries, products are manufactured and sold to any one requiring the same. In construction industry, there has to be a definite order for the construction from a user or a customer before the work begins.

3. Each structure is built *in-situ* :

- The Civil Engineering structure is constructed on the location where it is to be used. The motor car manufacturer assembles his motors within his factory and then sells to the consumers.

- The Civil Engineering on the other hand must take his factors of production to the promoter's chosen location for the project and that is where he must undertake the physical task of construction.

- Pre-fabrication techniques have an influence upon the in-situ aspect. Pre-fabricated slabs, beams etc., are manufactured now-a-days in factories such as siporex and used when necessary. It substantially reduces the cost in site operation but does not diminish the in-situ characteristics.

1.5.2 Civil Engineering Works

- Civil Engineering works are of various types and of different natures. It consists of the following types :
    1. Extraction of engineering materials such as soil, murum, sand, limestone etc.
    2. Manufacture of engineering materials such as bricks, tiles, pipes, cement, sheets etc.
    3. Construction of buildings, roads, railways, bridges, dams, harbours, airports, water supply, sewage, irrigation, hydro-electric power schemes etc.

- Any of the above works are not like production of a article or machinery from a factory or industry where there is set technique of production with controlled conditions (indoor nature). But site condition of civil engineering project differ invariably.

1.6 CLASSIFICATION OF CONSTRUCTION WORKS

Construction works are broadly classified as follows :

(1) Light construction, (2) Heavy construction, (3) Industrial construction.

1. Light Construction :

It includes works with light structural members and light foundation. Examples are residential, educational, commercial and other types of buildings, village and city roads, small water supply and sewage works and light industrial sheds etc.

2. Heavy Construction :

It includes heavy structural members on massive foundations requiring heavy machinery, equipment and large quantities of material, labour and finance for their construction.

Examples : Canals, Highways, Bridges, Railways, Irrigation Project, Docks and Harbours.

3. Industrial Construction :

It includes construction work related to industries involving specialised equipment and intricate installations, requiring special know-how and skill.

Examples : Steel mills, Oil refineries, Fertilisers, Chemical plants and Atomic reactors.

1.7 AGENCIES ASSOCIATED WITH CIVIL ENGINEERING CONSTRUCTION WORK

Civil Engineering Agencies : There are various Civil Engineering agencies contributing their share in executing civil engineering structures such as - (1) Users owners, (2) Promoters, (3) Financers, (4) Architects, (5) Consulting engineers, (6) Designers, (7) Builders/Contractors etc.

1. Users/Owners :

- One who uses the Civil Engineering work is user. Everyone is using civil engineering work and therefore, everyone is a user. Civil Engineering works may be buildings, roads, railway-lines or any ether type of civil engineering projects.

2. Promoters :

- A person/firm, who initiates Civil Engineering work is the promoter. Promoter may be a single individual or co-operative society or private firm or Ltd. company or government or semi-government agency.

- Promoters initiate the work, order the commencement of work and get the work executed through various agencies. Promoter is the most important link in the whole contract system because it is he, who ultimately employees everybody including consulting engineer upto the tea-boy. Without a promoter, no buildings or public works be ever constructed. Promoter is the source of inspiration responsible for the construction of Civil Engineering works and it is the promoter's wishes that consulting engineer interpretes in terms of actual construction.

Types of Promoters :

1. Government Department : Public works and Housing, Irrigation and Power Department.
2. Semi-Government : M.S.E.B., C.I.D.C.O., Zilla Parishad, Environmental Engineering Department etc.
3. Co-operative Societies.
4. Individual.

Scope :

- Promoters have : (1) to initiate the project, (2) to administrate the project, (3) to exercise over-all control.

Functions of Promoters :

1. To furnish full details and requirements, in the beginning, to the Architect or consulting engineer to complete the work in time and in given budgetory grants.
2. To appoint suitable agencies for planning and execution of the work.

3. To arrange for the finance.

4. To make payments to various agencies.

5. To adhere to all the requirements of relevant laws, technical as well as general rules, regulations and procedure.

6. To take all the responsibilities of getting the work done and to fulfill all the requirements as promised.

7. To get the work done through various agencies as per drawings specifications and to their entire satisfaction.

3. Financers :

- Financers are the persons or organizations who make investment in the project, to be ultimately recovered with profit from the users/promoters in suitable number of installments.

Types of Financers :

1. Private Financers
2. Public Financers

1. Private Financers : These are persons or companies who finance the project with the only aim of earning their profit. They are not concerned with the purpose, utility and the success of the project.

2. Public Financers : These are the organizations who finance only those projects which are in the interest of general public.

Examples of Public Financer : (1) Scheduled Banks, (2) Land Development Banks, (3) Maharashtra Industrial Development Corporation, (4) Housing Boards and (5) World Bank.

Scope :

- They have to provide the finance as agreed and to ensure the completion of all formalities for giving as well as recovering the loan.

Functions :

1. To ascertain the feasibility of the project before any loan is sanctioned.
2. To decide the procedure of sanctioning and recovering the loan.
3. To prepare and maintain all the documents and records as required.
4. To complete the required legal formalities.

4. Architects :

- An Architect is a person who is qualified in architecture and is recognised by various statutory bodies like Municipality, Corporation and trust etc. to practice in civil engineering works where space, elevation and appearance are predominant factors of construction.

- On contracts of such works the link between the owner and the contractor is the architect. An architect deals mainly with houses, public buildings, townships etc. which require proper layout planning and appearance. His scope, duties and functions are practically same as that of consulting engineer.

5. Consulting Engineer :

- Consulting Engineer is the mediator between the promoter (who wishes to construct Civil Engineering work) and the contractor (who constructs the Civil Engineering work).

- He is technically qualified person and is recognised by various statutory bodies like Municipalities, Corporation, Trusts etc., to practice in Civil` Engineering to the completion of the project including planning, designing and supervision of the construction, issuing completion certificate.

- Consulting engineer generally does advisory job in civil engineering project where engineering features are more predominant such as Dams, Bridges, Building Complex etc.

- He is a chartered engineer who, because of his technical qualifications, experience, is competent to advise, in engineering works.

Scope :

- To advise promoter and the contractor for satisfactorily completion of the work, right from preparation of preliminary scheme to completion certificate. He has to do mainly consulting job so that promoter gets good value of his investment and contractor gets good return of his efforts.

Functions of Consulting Engineer :

1. To advise the promoter/owner for the preparation of tile preliminary scheme.
2. To prepare detailed drawings, estimate, specifications, designs etc.
3. To obtain approval for the work from statutory bodies.
4. To advise the promoter/owner for selection of tender and fixing the agency for construction.
5. To provide working details to the agency and to give clear picture of the work to be completed.
6. To supervise the work, guide the contractor and his workers.
7. To take the measurements of work completed and certify the payments.
8. To establish the proper harmony, co-ordination among the various agencies involved in the project.
9. To give the decisions on professional judgement remaining impartial in all respects.
10. To act as an arbitrator to settle the disputes on the mutual assent of both parties, owner and contractor.
11. To protect the client against the dishonest contractor or the contractor against the unfair client.

6. Designers :

- Designers are the persons who are specialist in designing complicated components of a project such as R.C.C. designers, water filter designers, and waste disposal experts.
- They charge suitable fee for their services. Designers are generally referred by Architects/Consulting Engineers.

Scope :

- They have to design the elements of the project which needs special expertise.

Functions :

1. To provide complete design, drawings specifications etc., so as to enable the Architect/Consulting engineer to execute the same work without any difficulty.
2. To inspect and check the work for proper, implementation of design to ensure the standard work.
3. To help and guide the site engineer in all the aspects for correct, speedy and economical execution of works as per design.
4. To conduct required tests after completion of the component so as to be sure about successful implementation of the design.

7. Contractors/Builders :

- Contractors/Builders are the persons who actually execute the work according to the plans, designs and specifications, and as per the terms and conditions of the contract. Contractor is selected by inviting tenders publicly and offering a contract to the lowest bidder. They have to execute the work as per the direction of the Architect/Consulting engineer.

Scope :

- They have to execute the work as per contract agreement under the direction of the site engineer for which detailed specifications are stipulated.

Functions :

1. To maintain a suitable organization required for carrying out the work.
2. To arrange for necessary tools, plants, equipment and machinery required for the work.

3. To obtain registration from bodies such as P.W.D., etc. so as to be entitled for undertaking contracts of a particular work for a particular amount.

4. To arrange for required materials including its necessary testing.

5. To arrange for necessary labour required for the work.

6. To ensure safe working conditions and to provide fair wages and required amenities to the workers.

7. To arrange the schedule of work in consultation with the site engineer and carry out the work as per programme and to ensure the completion of work as far as possible in time-limit.

8. To take the full responsibility of the work till its completion and handing over.

9. To rectify any defect observed in defect liability period.

10. To observe the rules and regulations in force in obtaining, storing and using material requirement for the work.

1.7.1 Two Broad Categories of Agencies

The Civil Engineering agencies are broadly divided into two groups :

1. Planners,

2. Builders/Contractors.

1. Planners :

The planners are those who work out and prepare the details of a project such as drawings, design, specifications and estimates etc.

These details are essential for arranging the required finance, materials, labour, machinery etc. These details are also required for inviting tenders and fixing the agency for execution of work. These details are also very useful to the builder at the time of actual execution of work.

Planners are :

1. Government Departments : P.W.D., Irrigation and Power Department.

2. Semi-Government : M.S.E.B., C.I.D.C.O., Z.P., Environmental Engineering, MIDC etc.

3. Private Consulting Agencies/Firms.

2. Builders/Contractors :

*   The builders/contractors are those who actually execute a project as per planning and terms of agreement. They have to arrange for the required labour, material, tools and plants, equipment and machinery.

*   Initially they have to make financial commitments. They get periodical payments from the user/promoter/planner as per the terms and conditions of contract-agreements for the work completed.

Examples of Builders :

1. Government Departments : P.W.D. Departmental work.

2. Semi-Government Departments : M.S.E.B., Z.P., Environmental Engineering etc.

3. Private Contractors - (1) Gammon India Ltd., (2) Hindustan Construction Co. Ltd. etc.

Co–ordination of Various Agencies :

Co-ordination between various agencies is required :

1. To concentrate group efforts to achieve the common objective.

2. To improve the efficiency in business.

3. For survival of the organization.

4. To create good effects upon the general level of moral in an organization, sense of responsibility and faithful working.

5. To share the credit of work.

1.8 STAGES IN CONSTRUCTION

The various stages in construction are :

1. Initiation of proposal.

2. Acceptance of proposal in principle by competent authority.

3. Survey, Planning, Design, Drawing and preparation of estimate.

4. According of administrative approval and allotment of funds.

5. Preparation of tender documents and acquisition of land and inviting tenders.

6. Preparation of comparative statement and allotment of work.

7. Execution of work.

8. Completion Report Part-I (submitted on the completion of the work/project).

9. Completion Report Part-II (closing of accounts after disposal of surplus stores).

Broadly speaking the stages of construction are :

1. Pre-tender stage, and

2. Post-tender stage.

I.   Pre-tender Stage :

The following activities are included in pretender stage :

1. Selection of site for construction and finalising the outline of the works to be constructed.

2. Acquisition of land.

3. Soil investigations, preparation of estimate and designs. Preparation of tender.

4. Obtaining administrative approval and allotment of funds.

5. Arrangements for resource mobilization i.e. labour force, stores, equipments and finance.

6. Technical sanction for safety of structure.

II. Post-tender Stage :

1. Contract and planning for execution.

2. Execution of works according to the contract and supervision.

3. Inspection of works including network inspection.

    Planning and other methods to ensure completion of the project within the stipulated time.

4. Finalisation of accounts :

    (i) Completion of the work/project.

    (ii) Closing of accounts after disposal of surplus stores.

5. Maintenance.

1.9 RESOURCES OF CONSTRUCTION

Resources in a construction project are made up of :

1. Plant equipment and machinery required for the project, land and buildings.

2. Construction stores such as cement, bricks, steel, timber, aggregates and other fixtures and fittings etc.

3. Skilled and unskilled manpower.

4. Supervisory staff consisting of technicians and technocrats.

(Technician concerned with the execution and technocrats concerned with administration, policy making).

- In addition to the above resources, administrative cell is required to exercise control and command. The management is responsible for making best use of all resources.
- The management is responsible for combining the resources to achieve the greatest productivity.

1.10 JOB LAYOUT/SITE LAYOUT

- Job layout or site layout is a scaled drawing of the proposed construction site showing the relevant features such as :

    1. Entry point at site.
    2. Exit point from site.
    3. Office.
    4. Security guard room/cabin.
    5. Storage space for materials like cement.
    6. Stack location for sand, brick, reinforcement.
    7. Washrooms/toilets.
    8. Bar bending area.
    9. Space of machines like concrete mixers.
    10. Area for keeping equipments.
    11. Labour housing.

- The site layout for various construction activities consists of :

    (a) Access road for lories, trucks, cars etc.

    (b) Sheds for storage of material such as cement, lime, timber machines etc.

1.10.1 Preparing Job/Site Layout

- A plan in which the arrangements for placing site office, store room, labour quarter, medical aid centre, godowns for keeping construction materials and other facilities are properly prepared or chalkout, is called as job layout or site layout.
- The arrangements for processes should be such that the work is done smoothly and in orderly manner.
- In the job layout plan, there should be proper co-relation and co-ordination among the different units, in such a way that, site office and warehouse are placed close to the entrance of the site so as to have a better contact to the visitors.
- The areas should be properly alloted in such a way that time required in carrying materials is minimum, which reflects on increase in efficiency.
- **Job layout depends upon following factors, namely,**

    1. Location, area and topography of the site,
    2. Method of construction,
    3. Nature and type of work,
    4. Requirements of site office, store room, labour quarter, godowns, first aid,
    5. Availability of space,
    6. Availability of resources.

1.10.2 Site Clearance

- Site clearance is very important factor to be considered in the job layout plans, before commencement of project work.
- There should not be any obstruction like trees, plants, shrubs, bigger size stones etc.

- Approach road should be wide so as to allow the truck for loading and unloading the constructional material. If site ground is uneven, then it should be made plane. For this, contouring is done in advance.

- In short, site should be cleared by all the way, otherwise no activity can be launched in construction processes.

1.10.3 Purpose/Necessity of Job Layout

1. To save the time.

2. To safeguard construction material.

3. To store construction material.

4. To avoid the cost of cartage.

5. To avoid the in-time delay in getting the material.

6. To use the equipment/machinery and to save the time and cost of project.

7. To provide safety to workers.

8. To reduce the cost of project.

1.10.4 Advantages of Job Layout

1. Save the time of project.

2. Save the cost of project.

3. Smooth working of the project.

4. Safety in working.

5. Proper use of materials.

6. Economical use of equipments/machinery.

7. To minimise the wastage of material.

8. To protect the material from damage.

1.11 LAYOUT PLAN FOR STORAGE MATERIALS

- Before starting the construction work, a layout plan is prepared.

- The plan gives proper location for machines, proper and safe storage for materials and labours/men to facilitate the work as per schedule.

1.11.1 Important Points for Preparation of Plan for Job Layout

1. Materials are required to store near the work to reduce the cost of transportation and easy handling of material.

2. Machine are placed in the most advantageous positions.

3. For administrative control office location should be at proper place.

4. Proper access to vehicles coming at site for various reasons.

5. Sufficient place for movement should be available near the work for easy movement of labour and machinery.

6. Proper and suitable locations/places should be provided to store additional/excess materials.

7. The layout should be planned in such a way that this facilitates maximum efficiency in minimum movement, to save time and money.

8. Fencing and hording at site is required for safety and identification of project.

1.11.2 Temporary Services at Site

Temporary services at site are : 1. Water supply. 2. Electric supply. 3. Telephone lines.

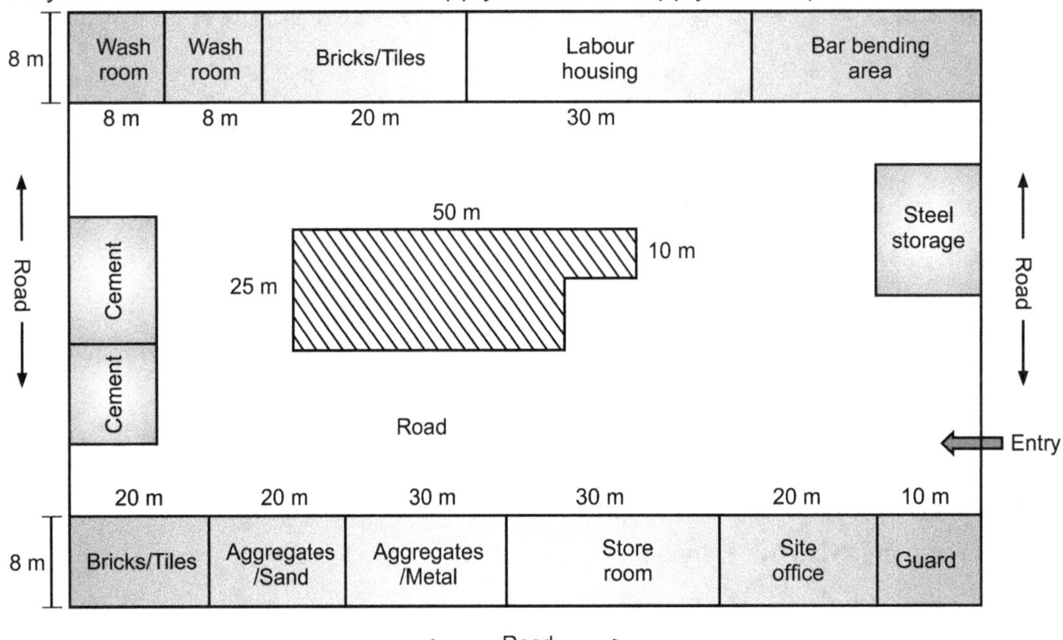

Fig. 1.1 : Job layout for a multistoryed building

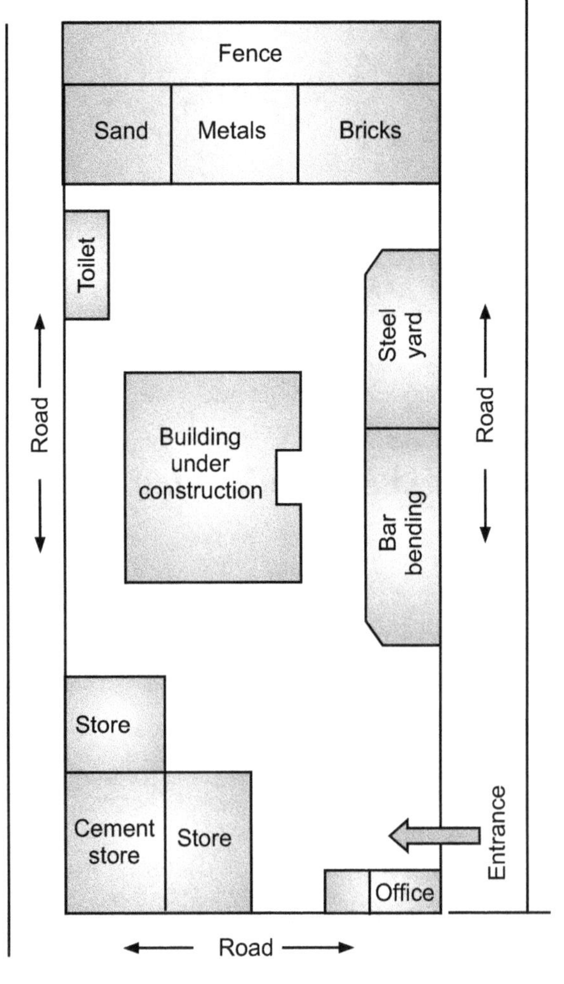

Fig. 1.2 : Site layout for multistoryed building

Important Points

- Organization means 'a group of people working together to achieve a common goal'. A common goal in contractors organization is to complete the work as per tender documents as early as possible and to achieve maximum profit. A common goal in a factory may be to have maximum production.

- **Basic Activities in Organization :**

  (1) Technical activities   --   Production, manufacture etc.

  (2) Commercial activities   --   Buying, selling

  (3) Financial activities   --   Optimum use of capital

  (4) Security activities   --   Protection of property and persons

  (5) Accounting activities   --   Stock takings, preparation of balance sheets etc.

  (6) Administrative activities   --   Planning, organizing, controlling.

- **Types of Organization :**

  (i)   Military or line type organization.

  (ii) Line and staff organization.

  (iii) Functional organization.

- **The main branches of P.W.D.** are :

  (1) P.W.D. (Public Works and Housing Department).

  (2) P.W.D. Irrigation.

  (3) Environmental Engineering Department.

- **Types of Contracting Firms :** The following are important contracting firms :

  (i)   Sole proprietorship,

  (ii) Partnership,

  (iii) Private limited company,

  (iv) Co-operative societies.

- One of the most distinguishing feature is the ever changing labour force. More often and then, not a new labour force is employed for every new work taken up. On completion of a project or work, the labour shifts to a new place or site in search of work.

- **Personnel Management :** For the efficient use of all other resources, efficient use of human resources in employment is essential otherwise raw materials will remain stocked, machines will remain idle and money will be blocked up and work will come to a standstill. Hence, management of man power has greater importance.

- **Type of Labour :** The following types of labour are generally found working in construction industry :

  (1) Unskilled labour.

  (2) Semi-skilled labour.

  (3) Skilled labour.

- India has attained her political freedom, however, her economical freedom is yet to be achieved. Statements are formulating plans and projects - mainly Civil Engineering projects such as dams, canals, roads, railways, bridges, buildings etc., for the development of the country.

- But these projects should be completed effectively, satisfactorily and economically by Civil Engineering industry to help to achieve economical freedom.

- **Nature of Civil Engineering Management :** Civil Engineering is defined as "the art of directing the great sources of power in nature for the use of and convenience of man". For many centuries, the art of Civil Engineering is used by a man for design and construction of dams, roads, bridges, buildings etc. Civil Engineers are applying their skills to raise the general standards of living of the human beings.

- **Types of Promoters :**

    1. Government Department : Public works and Housing, Irrigation and Power Department.

    2. Semi-Government : M.S.E.B., C.I.D.C.O., Zilla Parishad, Environmental Engineering Department etc.

    3. Co-operative Societies.

    4. Individual.

- **Builders/Contractors :** The builders/contractors are those who actually execute a project as per planning and terms of agreement. They have to arrange for the required labour, material, tools and plants, equipment and machinery.

Practice Questions

1. Explain 'organization' by Georg R. Terry.
2. Enlist basic activities in organization.
3. Enlist type of organization.
4. Draw flow chart of :
    (i) Line type organization.
    (ii) Line and staff organization.
5. Describe organization of P.W.D.
6. Give the function of :
    (i) Chief enginee.
    (ii) Superintending engineer.
    (iii) SDO.
    (iv) Junior engineer.
7. Enlist types of contracting firm.
8. Give advantages of partnership firms.
9. Describe public limited company.
10. Enlist and describe principle of organization.
11. Define personal management and give its functions.
12. Describe types of labour in construction industry.
13. Emphasise the importance of construction industry in the National Development.
14. State the special characteristics of Civil Engineering work.
15. State the agencies associated with Civil Engineering construction works. Give two functions of each.
16. Name any four Civil Engineering construction small scale units.
17. State the advantages of small scale industries in a developing country like India.
18. State the types of promoters, give the functions of promoters.
19. Mention the duties of consulting engineer.
20. State the functions played by the following in construction industry : (1) Designer, (2) Architect, (3) Financer, (4) Contractor.
21. State the stages of works in construction. Enlist the works to be carried out in each stage.
22. Describe the role of construction activity in national development.
23. Describe 'preparation of job layout'.

PLANNING AND SHEDULING

Syllabus

2.1 Identifying broad activities in construction and allotting time to it based on rate analysis, Methods of scheduling, Development of bar charts, Merits and Limitations of bar chart.

2.2 Elements of Network : Event, Activity, Dummy activities, Precautions in drawing network, Numbering the events.

2.3 CPM networks, Activity time estimate, Event times by forward pass and backward pass calculation, Start and Finish time of activity, Project duration, Floats, Types of floats - Free, Independent and Total floats, Critical activities and Critical path.

2.4 Purpose of crashing a network. Normal time and Normal cost, Crash time and Crash cost, Cost slope, Optimization of cost and duration.

2.5 Material management - Ordering cost, Inventory carrying cost, EOQ.

2.6 Store management : Various records related to store management, Inventory control by ABC technique.

Objectives

After learning this chapter, student will be able to

- Draw the bar chart for the given construction project.

- Draw the network for the given construction project.

- Compute activity times, event times, floats and duration of the given construction project.

- Calculate optimum cost and duration of the given project.

- Carry out resource levelling for the given project.

- Calculate EOQ in the given situation.

- Identify the ABC analysis for the given items of store.

- Identify the forms pertaining to the given store item with justification.

2.1 PLANNING

- The aims of construction management are to execute the work in efficient manner as per drawing, design and specification within the prescribed time-limit and with the maximum possible economy in expenditure.

- For efficient execution of the work it is essential that the scheme of working should be first decided. It means deciding what to do, when to do and how to do and this is known as planning. It involves the determination and formulation of steps and processes for doing a particular work at particular time.

2.2 NECESSITY AND IMPORTANCE OF PLANNING

- The civil engineering works executed without planning may result in considerable delays in completion, higher expenditure than estimated and lesser benefits than anticipated. All these cause over capitalization, seriously affecting the national economy and retarding the planned development of the country.

- Planning should be done for all the works. It needs lot of experience and devotion of much time and thought in order to attain best programming. The planner relies on sound judgement based on knowledge, experience and mathematical methods.

- Various possible alternative methods are worked out and finally one is decided considering all the aspects of work such as requirements of labour, material, machines, finance, arrangements to be made for the provision of construction facilities, likely uncertainties in execution due to various unpredictable factors.

2.2.1 Controlling

- Controlling means checking constantly the progress of work with the planned programme and identifying areas of deficiency, if any, so that remedial steps can be taken.

Functions of Controlling :

1. Keeping a watch over the progress of work.
2. Controlling the quality of work.
3. Controlling the use of machines, materials, output of labour to avoid wastage.
4. Controlling the expenditure.

2.2.2 Planning at Different Stages

- Planning is done at different levels, in different fields and at different stages as per size and nature of the project, e.g.
    1. Planning at National level.
    2. Planning at State level.
    3. Planning at Promoter's level.
    4. Planning at Contractor's level.
    5. Administrative planning.
    6. Technical planning.
    7. Job planning
        (i) Pre-tender planning,
        (ii) Post-tender planning/Contract planning.

- Job planning i.e. Pre-tender planning and Post-tender planning are described below :

(i) Pre-Tender Planning :

- Pre-tender planning is the planning undertaken by the tenderer after receipt of tender notice and before submitting a tender.

- It helps to the tenderer to make a proper bid, keeps him ready for taking up the work. It includes :
    1. Careful study of drawings, specifications, conditions of contracts, time-limit.
    2. Working out quantities of items of work, requirements of materials, labour, machineries.
    3. Facilities for camp.

(ii) Post-Tender Planning :

- After getting the contract, the contractor has to undertake further intensive planning, this planning is known as Post-Tender planning or Contract planning. It includes :
    1. Studying alternative methods of construction and deciding suitable one.
    2. Deciding about sub-contracting.
    3. Planning the location of site, details of camping, office service load, layout of site, labour camps, facilities to workers and staff, transport etc.
    4. Preparing various schedules, with dates for the inceptions and completion of each item of work.
    5. Safety measures to avoid accidents.

2.2.3 Planning Methods

- The generally accepted methods of planning and controlling of Civil Engineering works are :

1. **Bar Charts :**

   (a) Gantt Bar Chart.

   (b) Mile Stone Chart.

2. **Analysis of Network :**

   (a) Critical Path Method.

   (b) Programme Evaluation and Review Technique.

2.2.4 Scheduling

- Scheduling is a graphical device which shows series of jobs to be performed with dates of starting and completing each job or operation as well as sequential relationship among the various jobs.

- For preparing construction schedule following procedure is generally adopted :

   1. Breaking down of work into series of activities (operations) for phasing.

   2. Sequencing the activities.

   3. Planning the activities within time.

- While preparing schedule following factors are taken into account :

   1. Quantity of each activity, rate of working and time required to complete.

   2. Difficulties in procurement of material, time for delivering materials etc.

   3. The type, quantities, duration of machines and equipments required.

   4. The type of labour, number of labour and period during which they will be required.

   5. Suitable margin of time for unexpected delays due to bad weather, accidents, earthquake, floods etc.

Methods of Scheduling :

   (i) Bar charts or Gantt charts.

   (ii) Milestone charts.

   (iii) Network analysis.

2.3 PLANNING BY OWNERS SIDE AND CONTRACTORS SIDE

(1) Planning by owner :

- A civil engineering project should begin with a though investigation of its scope and feasibility. The owner should methodically plan the following activities :

   (a) Ideas originated by individuals are studied with regard to cost and benefits so as to establish the economic viability or social utility of the project and to arrange adequate funds for the project.

   (b) To appoint a project steering committee if necessary.

   (c) To appoint a project manager who will have continuing responsibility to the owner throughout the construction process.

   (d) To carry out extensive investigations which will include both technical and non-technical investigations.

   (e) To carry out market survey for resource identifications.

   (f) To study various alternatives and identify the most feasible one.

   (g) To prepare the project report.

   (h) A realistic and detailed cost estimate of the project shall be prepared.

   (i) To prepare working drawing, specification and all arrangement for inviting tenders and to appoint a suitable contractor.

(2) Planning by contractors side :

- The contractor executing the civil engineering construction works has to plan the following important activities so as to achieve the completion of the project as per schedule and to derive adequate benefit.

    (a) Construction planning includes the preparation of :

    (i) Construction schedule.

    (ii) Manpower schedule.

    (iii) Plant and equipment schedule.

    (iv) Material delivery schedule.

    (b) To control site operations such as :

    (i) Temporary and permanent work.

    (ii) Supply of materials and equipment.

    (iii) Co–ordination of various sections.

    (iv) Supervision for quality control.

    The construction work has to be carried out in a planned manner to prevent wastage of manpower, material and money and to avoid disruption of project schedule.

    (c) To prepare records of the actual construction work. The changes are also recorded for reasons of technical performance and financial implications.

    (d) Every activity must be planned well in advance so that training and recruitment of staff and deliveries of equipment match the construction schedule.

    (e) To submit running bills to the owner for payment based on the progress of work and materials brought at site.

    (f) To keep proper interaction between the owner, engineer, architect for smooth and efficient execution of a constructions project.

- Proper understanding of functions/activities of each team plays a vital role in achieving speed, economy, efficiency and quality in all construction projects.

2.3.1 Study of Drawings and Designs

- In the Designing stage of the construction project, detailed design, working drawings, specification, bill of quantities, final cost estimate are prepared. The purpose includes the preparation of working drawing specifications and all arrangements for inviting tenders.

2.3.2 Resources for Execution of Works

- The main resources needed for the execution of construction works are :

(i) Materials : Materials such as bricks, stones, cement, aggregate, steel, shuttering, scaffolding, timber, water supply, sanitary and electrical fittings, petrol, oil, lubricants etc. are required for construction.

(ii) Manpower : Manpower in the form of technical and managerial personnel and work force in various trades is essential to carry out construction activities. The technical and managerial personnel are essential for efficient use of human resources and to achieve project completion within estimated time and budget. Technical personnel include engineers, architects, quantity surveyors, supervisors, technicians etc. The work force consists of skilled and unskilled workers.

(iii) Machinery : For any construction work, various plant, equipment and tools are required e.g. batching plant, mixers, trucks, tractors, excavators, dumpers, cranes, pumps, generators, workshop equipment etc. as per demand of the construction work. For efficient construction activity these plant equipment need to be properly maintained.

Power is essential resource required for lighting, running the plant and equipment and other facilities.

(vi) Funds : Adequate funds should be available for smooth implementation of the project. Funds form an important resource.

Financial planning is essential for smooth cash in-flow and out-flow to avoid delays in construction activities. All other resources are dependent on the availability of funds.

(v) Space : Space is essential for efficient execution of works. It is mainly :

(a) Storing material.

(b) Providing yards for bar benders, carpenters, installation of equipment and plant, repair work shops, casting yards etc. depending on the type of jobs.

(c) Site office, labour camp etc.

2.4 GANTT BAR CHART

- It is the graphical representation of various activities of the work. In Gantt chart, time in days or weeks is marked along the horizontal axis and the activities are represented along the vertical axis. The programme of various activities is represented by horizontal bars indicating duration of execution, and hence it is known as bar chart.

- In 1899, Henry Gantt exhibited bars of various activities in the form of a chain and then developed a method known as Gantt Bar Chart.

- All the activities of work are listed on left hand side of the chart and proposed duration of their execution is represented by horizontal bars in front of the activities, in the right hand side of the chart. During the course of execution, work executed is marked by hatching the portion of the horizontal bars of respective activities and thus it compares actual performance with planned works over time, this hatched portion represents percentage of work done. Hence, this chart is also known as time and progress chart.

- This chart is popular with construction firms of small size.

- The progress of work is to be shown. day-to-day in the chart and chart is to be made up-to-date. This constant progress making on the chart is troublesome and difficult. This difficulty is now overcome by Mechanical Bar Charts which are made up-to-date mechanically.

Advantages of Gantt Bar Chart :

1. It is simple to prepare and interpret.
2. Each activity is shown separately.
3. Modification can be done easily without disturbing other bars.
4. Used as a visual aid.

Disadvantages/Limitations of Gantt Bar Chart :

1. The sequence of activities is not clear.
2. Interdependence of activities is not clear.
3. Chart does not give the peak rate of work necessary for timely completion of work.
4. Chart does not give the overall progress, hence it is not possible to review or revise the programme.
5. Chart can not be used for quick evaluation of alternatives.
6. Chart can not be used for effective controlling.
7. Chart does not give the optimum duration of the project.
8. Chart does not incorporate uncertainties or degree of tolerance for delay in the estimated duration of activity.

Example of Gantt Bar Chart :

SOLVED PROBLEMS

Problem 2.1 : *Draw the Gantt Chart or Bar-Chart for a simple project including following activities.*

Activity A and B can start simultaneously and proceed parallel.

Activity A is to be completed in 24 weeks and B is to be completed in 20 weeks.

Activity C starts 4 weeks after the start of activity B and takes only 16 weeks.

Activity D has to proceed activity B and takes 6 weeks.

(i) What is the total project duration.

(ii) If the progress of activities are as scheduled shown the progress at the end of 12th week on the chart.

Solution : The Bar-chart is drawn below.

The progress at the end of 12th week is shown hatched in the chart. The total project duration is 26 weeks.

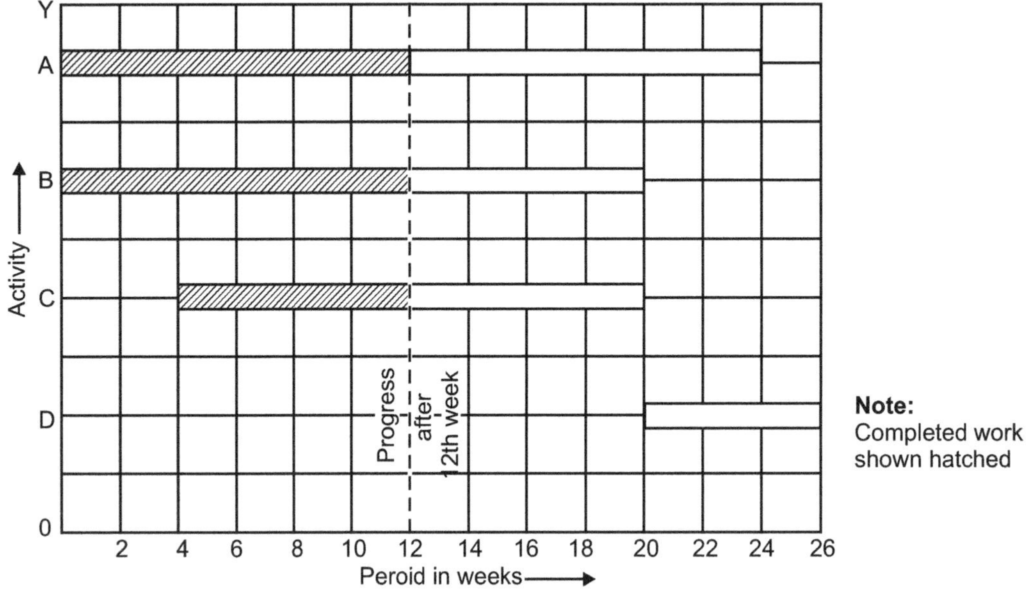

Fig. 2.1 : Grant bar-chart

Problem 2.2 : *Draw the Bar Chart for the following project. The project begins on Wednesday, November 15th with five work days in a week. Draw the Bar Chart with the horizontal scale denoting calendar dates.*

Activity	Days	Activity	Days
1	8	5	3
2	4	6	8
3	7	7	12
4	9	8	15

Activity 1 and 2 can occur concurrently. Activity 3 can take place after activity one is completed. Activity 4, 6 and 3 can occur concurrently. Activity 8 can start 4 days after the commencement of activity 6. Activity 7 should follow activity 5. Activity 5 can be concurrently with activity 8.

Answer the following :

(i) On what calendar date can we can expect the project to be completed ?

(ii) On December 3rd what is the progress ?

Solution : The Bar chart is drawn below.

The project begins on Wednesday, November 15th. Saturday and Sunday are holidays. The total Project duration is 27 days.

The progress on December 3rd is shown hatched.

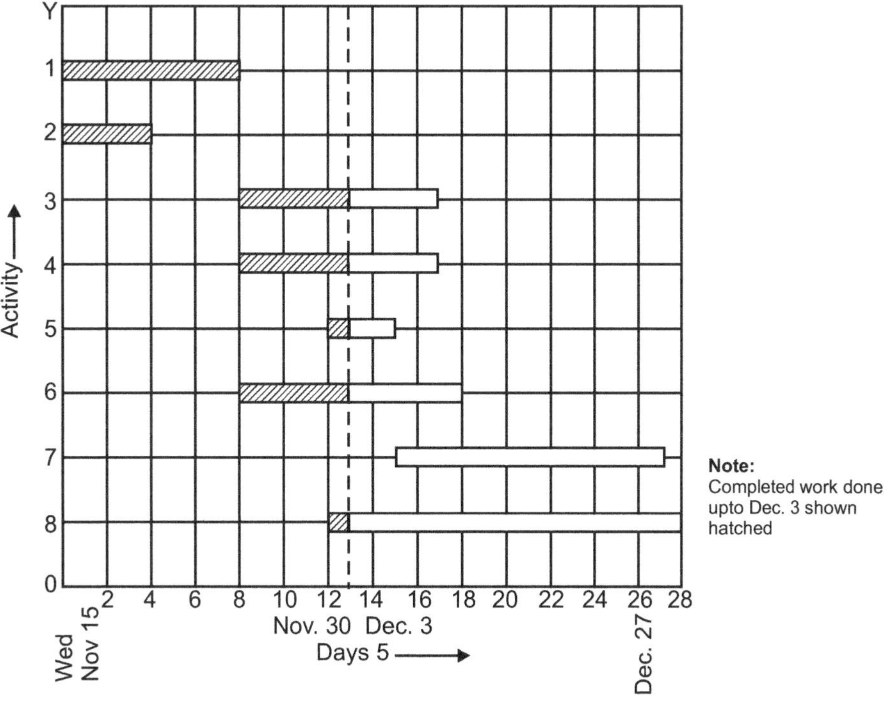

Fig. 2.2

2.5 TERMS USED IN THE NETWORK

1. **Network :** It is a graphic representation of the entire project in terms of its activities through the use of arrows and nodes showing their inter-relationship.

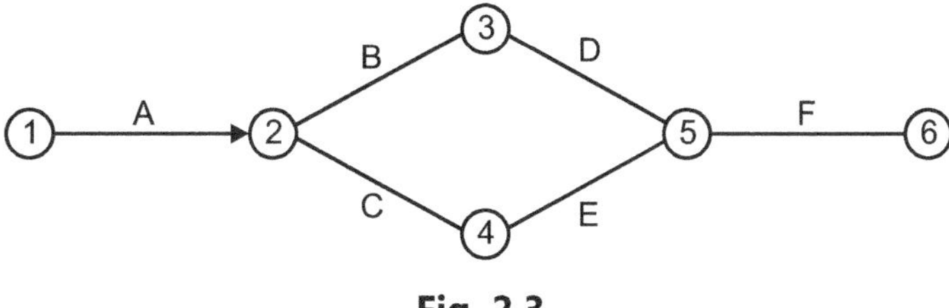

Fig. 2.3

2. **Activity :** An activity is a part of item or operation of the project. It is represented by an arrow on the network. The tail of the arrow shows the start of the activity and the head shows the end of the activity. The shows are now drawn to scales, length of the arrow has no significance. Each activity has its duration, (time required to complete the job) which is shown on the network. It consumes some resources.

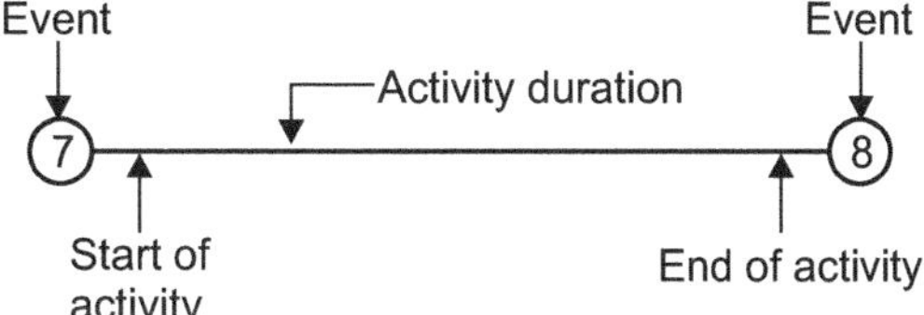

Fig. 2.4

3. **Event :** An event is a well defined point or a stage on the network just like a station on a railway line.

    All the previous activities, which are executed are marging in it. And the further activities which are still to be executed are bursting out from it i.e. it is either beginning or end of an activity. It does not consume any resources. It is represented by a circle. The events are numbered in their sequential order on the network.

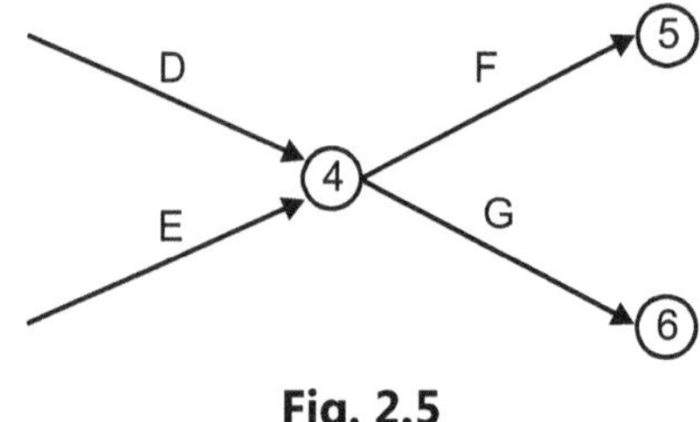

Fig. 2.5

4, 5, 6 are events.

D and E are marging activities⎫
F and G are brusting activities⎭   For event 4

4. **Activity duration (T) :** It is the anticipated time required to complete the given activity.

5. **Dummy activity :** It is fictitious activity with zero duration and no cost. It is used to maintain the sequential order of the activities in the network. It is shown by dotted line.

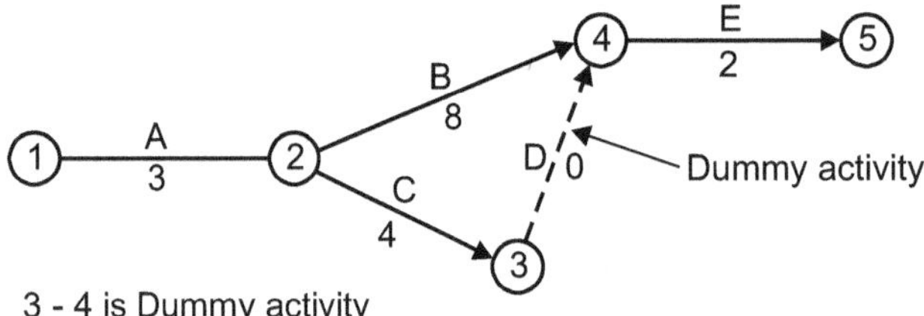

3 - 4 is Dummy activity

Fig. 2.6

6. **Restraint :** It is a restriction very similar to a dummy activity but has a duration which can either be negative or positive. It is used to fix intermediate dates with the network and thereby fix the relative start or finish of the parallel activities when these activities are not a coincident.

7. **Critical activity :** The activities lying on the critical path are called critical activities. It has a zero float.

8. **Critical path :** The path joining critical events is called the critical path of the network. The path given by critical activities is the critical path of the network.

9. **Critical events :** The events which have no float are the critical events. i.e. if $T_E = T_L$ then event is critical.

10. **Earliest Starting Time (EST) :** It is the earliest time by which an activity can start.

11. **Earliest Finishing Time (EFT) :** It is the earliest time by which an activity can be completed.

$$EFT \; = \; EST + T$$

12. **Latest Starting Time (LST) :** It is the latest time by which an activity must start so that the project is not delayed.

13. **Latest Finishing Time (LFT) :** It is the latest time by which an activity must be completed so that the project is not delayed.

14. **Earliest Event Occurrence Time (TE) :** It is the earliest time that the event can occur.

15. **Latest Allowable Event Occurrence Time (TL) :** It is the latest time by which the event can occur.

16. **Total Float (S) :** It is the difference between the maximum time allowed for an activity and its estimated duration. It is the amount of time by which the activity can be started late without delaying the project. It is also called total activity slack.

$$S \; = \; LST - EST = LFT - EFT = TL = EFT$$

17. **Free Float (FF) :** The free float of an activity is the amount of time by which the activity completion time can be delayed without interfering with the start of succeeding activities. Its use will not delay the completion of the project.

$$SE \; = \; TE = EFT$$

18. **Interfering Float :** It is the difference between the total float and free float. The use of it will delay the succeeding activity.

19. **Independent Float :** It is the difference between the total float and free float. The use of it will delay the succeeding activity.

20. **Numbering the Events (Fulkerson's Rule) :** Fulkerson's rule helps to number the events scientifically. It is used to maintain logical sequence.

    Step 1 : There will be single initial event in a network which has only arrows coming out of it. Number one (1) is given to this event.

    Step 2 : All arrows coming out of event (1) are neglected. This gives us with one or more initial events. These events are numbered as (2), (3), (4), (5) etc.

    Step 3 : Again neglect all the arrows coming out of these numbered events. A few more initial events may be created. These are also given number as per step 2.

    Step 4 : This sequence (operation) is continued until the last event of network is leached and numbered.

2.5.1 Developing Networks for A Work

Following procedure is followed in developing the network.

1. Project breakdown.
2. Network diagram.
3. Utility data with respect to time and cost.
4. Determination of critical path.
5. Activity times and float.
6. Scheduling.

1. Project Breakdown : The project is broken down into activities or jobs or operations, such as clearing site, excavation, P.C.C., stone masonry etc. The term activity is used to indicate each such job. Term event is used to denote completion of an activity. Thus, time is required to complete an activity, whereas event requires no time.

After preparing a list of all activities, the next step is to decide the relationship between all these activities and then arrange them in proper sequence, knowing preceding activity, concurrent activity and following activity. Thus, each activity is properly scrutinised, examined and properly placed in its order.

2. Network diagram : C.P.M. is a graph of operations. Each operation is represented by an arrow. Every arrow has a tail and a head. The tail of an arrow means the beginning of an operation and the head of an arrow marks the end of an operation.

Arrows can be used to express relationships among operations.

Every operation in a project is related to the other operations in the project in one or more of the following ways :

(a) It must precede some operations.

(b) It must follow some operations.

(c) It can be done at the same time when other operations are being performed.

The relationships between the arrows are indicated below :

(1) Operation A must precede operation B (B must follow A).

(2) Operation C can be done at the same time as operation D is being performed (C and D start at a common point).

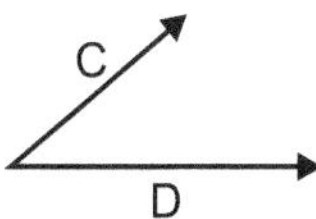

Now let us examine the following relationships among operations :

(a) A must precede B.

(b) E must precede F.

(c) B must follow A and E.

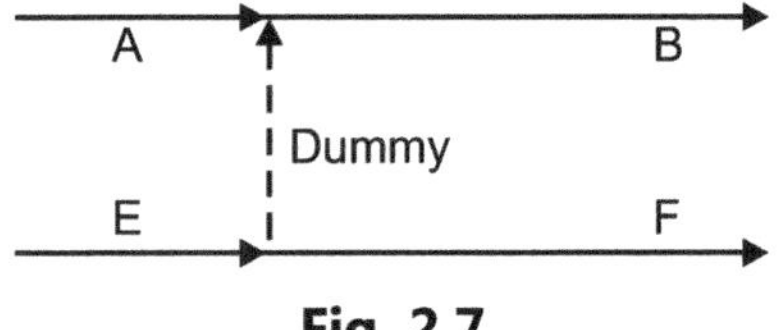

Dashed arrow is called a dummy operation.

Fig. 2.7

Table 2.1 : Common graph diagrams

(1)	**Logic :** Operation B can begin only after operation A is completed.
(2)	Neither operation B or C can start before operation A is completed, but B and C can be performed concurrently.
(3)	Operation C can begin after only operation A and B are completed.
(4)	Neither operation C or D can begin until both A and B are completed, but C can be started independent of D or vice-versa.

3. Utility Data with respect to Time and Cost : The direct cost of every activity is related to time of completion (Duration = T) of that activity by every possible method of carrying out that activity. This information is known as Utility Data. Utility data are analysed to know optimum cost and optimum time. This is beyond the scope of the syllabus. In our case duration of the activities will be directly given. Duration may be expressed in hours, days or weeks.

4. Determination of Critical Path : After drawing the network diagram and knowing the duration of all activities,

(1) Earliest Starting Time (EST)

(2) Earliest Finishing Time (EFT)

(3) Earliest Occurrence Time (TE) are computed by "Forward Pass' computation method and;

(4) Latest Starting Time (LST)

(5) Latest Finishing Time (LFT)

(6) Latest Occurrence Time (TL) are computed by "Backward Pass" computation method with the following equations :

$$EST = EFT \text{ of all event} \qquad \qquad \ldots (2.1)$$

$$EFT = EST + T \ (T = \text{Duration of activity}) \qquad \qquad \ldots (2.2)$$

$$LFT = LFT \text{ of head event} \qquad \qquad \ldots (2.3)$$

$$LST = LFT - T \qquad \qquad \ldots (2.4)$$

All these times are written on the network at proper places as shown.

Identification of Critical Path and Its Significance :

$$(7) \quad \overset{\displaystyle EST \qquad G \qquad EFT}{\underset{\displaystyle 16 \qquad \qquad 20}{\longrightarrow}} \quad (8)$$

$$(7) \quad \overset{\displaystyle LST \qquad T \qquad LFT}{\underset{\displaystyle 20 \qquad 4 \qquad 24}{\longrightarrow}} \quad (8)$$

Here, activity G is shown.

Its EST = 16, EFT = 20, LST = 20 and LFT = 24.

Activity G can be started on 17th day and can be finished on 20th day. But if this activity is delayed by one or two or three or four days, it will not hamper the programme because it can be completed upto 24th day without delaying the programme. This is an ordinary activity.

$$(11) \quad \overset{\displaystyle 32 \qquad J \qquad 36}{\underset{\displaystyle 32 \qquad 4 \qquad 36}{\longrightarrow}} \quad (13)$$

Here, activity J is shown

Its EST = 32, EFT = 36, LST = 32, and LFT = 36.

This activity can be completed, earliest on 32th day and latest on 36th day. If this activity is delayed even by one day it will delay the programme.

This is a critical activity. ($\because$ LST − EST = 0 or lFT − FET = 0)

In short, there are two figures giving EST and LST or EFT and LFT for every activity. The difference between these two figures gives the time available for delays and this is known as float. If there is no float or zero float, activity is known as critical activity and event as critical event. If the project is to be completed in time, these critical activities or critical events must be finished in time. The path joining these critical events is known as the critical path. It is shown by other colour or by thick lines.

5. Activity times and Floats : From the activity times, floats are worked out. Floats are of three types.

1. Total Float (T.F.)

2. Free Float (F.F.)

3. Interfering Float (I.F.)

$$T.F. = LFT - EFT \text{ or } LST - EST \qquad \qquad \ldots (2.5)$$

$$F.F. = EST \text{ of following activity} - EFT \text{ of activity} \qquad \qquad \ldots (2.6)$$

$$I.F. = T.F. - F.F. \qquad \qquad \ldots (2.7)$$

6. Scheduling : A Schedule is then prepared showing all the times and float calculations as shown in Table below. The information and data shown in schedule help a lot to the construction manager for planning and controlling the project.

Activity	Arrow	Duration	EST	LST	LFT	EFT	TF	FT	IF	Remarks

2.5.2 Applications of Network Technique to Simple Engineering Problems

The different applications of network techniques are as under :

1.   Civil engineering and construction projects including dams, bridge, buildings etc.

2.   Town planning.

3.   Ship building, construction and repair.

4.   Modification of existing plants for improvements.

5.   Research and development.

6.   Space programmes.

7.   Computer system.

Procedure for Developing a Network for a Construction Project :

1.   A list of activities included in a project, along with other depending activities is prepared.

2.   First a rough draft is prepared.

3.   The estimated time required for completing each activity is found out.

4.   This estimate is then written down on each activity.

5.   Scheduling computations are prepared and the earliest and latest allowable start and finish times for each activity is found so as to identify the critical path and to indicate the amount of slack on non-critical path.

6.   The final form to be used in the field is then prepared.

Advantages of Network over the Other Methods like Bar Charts :

1.   The sequence of operations is established.

2.   The inter-relationship between various activities is very clearly established.

3.   Planning and scheduling are distinguished clearly in the network.

4.   Important (critical) activities are stored out enabling the management for careful control.

2.6 CRITICAL PATH METHOD (ANALYSIS OF NETWORK)

- Bar charts are inadequate for planning large complex works, where high degree of control is necessary. In such works critical path method is used. When it is very necessary to adhere to the target dates of completion of work critical path method is generally used.

- The critical path method is a new powerful technique for planning and managing all types of construction projects. It is the representation of project plan by a systematic diagrams or network that describe the sequence and interrelation of all component parts of the project. This method is most suitable method for construction industry.

- The C.P.M. is developed by Morgan Walker of Du Pant, US.A. in 1957. The system is flexible, can be detailed to extent desired. Network analysis is amenable to computerization and hence even large, multipurpose projects of complex nature can be planned, scheduled and controlled with this system.

- The C.P.M. has proved to be useful and powerful tool in the hands of constructional engineers, as it provides the interlinking and determination of inter-dependency of various activities involved in the project. It provides a diagrammatical form of the project as a whole and from the study of C.P.M. chart it becomes easy to identify critical activities and further, the possible bottle-necks and difficulties likely to hamper the progress can be visualized well in advance and remedial measures may be taken to set the matter right in time.

2.6.1 Advantages of Critical Path Method

1. The entire project is viewed as a single entity, the project duration is known.
2. Critical activities, which affect the project duration are identified.
3. Interdependence of the various activities is very clearly established.
4. It ensures optimum use of men, materials and machines.
5. Various timings associated with activities, start, end, leeway available can be determined and can be used for controlling the project.
6. Method provides a very good tool also for cost control.
7. Manager gets valuable aid to assess the progress and can exercise the required control.
8. In case programme gets upsets, C.P.M. provides a tool for rescheduling and progressing the project.
9. Period of construction can be reduced, if necessary, by adopting crash programming.

SOLVED PROBLEMS

Problem 2.3 : *Draw the network and mark critical path for the following activities :*

Activity	Duration (days)
A, 1 - 2	3
B, 2 - 3	5
C, 2 - 4	4
D, 3 - 5	2
E, 4 - 5	4
F, 5 - 6	5

Solution : In this problem, project breakdown, sequence and duration is already given. So network can be prepared.

Steps : (i) Draw network as per sequence.

(ii) Number the events (1), (2), (3) etc.

(iii) Write symbol of an activity A, B, C etc.

(iv) Write duration of each activity,.

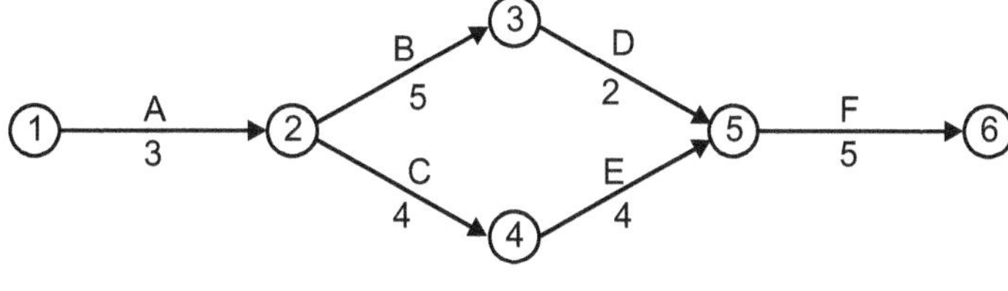

Fig. 2.8

Duration of all activities are known and now next operation is to compute the activity times.

Forward Pass :

For activity A, (1-2), it is a starting activity. It is to be started when time consumed is zero.

Earliest Starting Time (EST) of Activity A = 0.

Duration of activity A = 3 times.

It will be executed on 1st, 2nd and 3rd day of our work period.

It will be completed at the end of 3rd day.

Earliest Finishing Time (EFT) = EST + T (T = Duration = 3 days) = 0 + 3 = 3.

Next activities are B and C. They are to be started after completing activity A i.e. after completing 3 days i.e. after consuming 3 days period.

$$\text{EST of Activity B} = \text{EFT of Activity A} = 3$$

$$\text{EST of Activity C} = 3 = \text{EFT of Activity A which is preceding activity of B and C}$$

$$\text{EFT of Activity B} = \text{EST} + \text{T} = 3 + 5 = 8$$

$$\text{FET of Activity C} = \text{EST} + \text{T} = 3 + 4 = 7$$

Now, Activity D is to be started after completing activity B which is preceding activity of D.

$$\text{EST of Activity D} = \text{EFT of Activity B} = 8$$

$$\text{EFT of Activity D} = \text{EST} + \text{T} = 8 + 2 = 10$$

$$\text{EST of Activity E} = \text{EFT of preceding activity C} = 7$$

$$\text{EFT of Activity E} = \text{EST} + \text{T} = 7 + 4 = 11$$

Now, Activity F is to be started after completing is proceeding activity D and E. D Activity will be over at the end of 10th day. But E activity will be over the end of 11th day.

∴   Activity F will be started 11 days.

∴   EST of Activity F = EFT of Activity D or FET of activity E, whichever is maximum.

∴   EST of activity F = (Maximum of 10 and 11) = 11

$$\text{EFT of activity F} = \text{EST} + \text{T} = 11 + 5 = 16$$

All activities are over. We know EST and EFT of all activities. Now, write all times of the activities on the network at proper places (above the activity arrow-line).

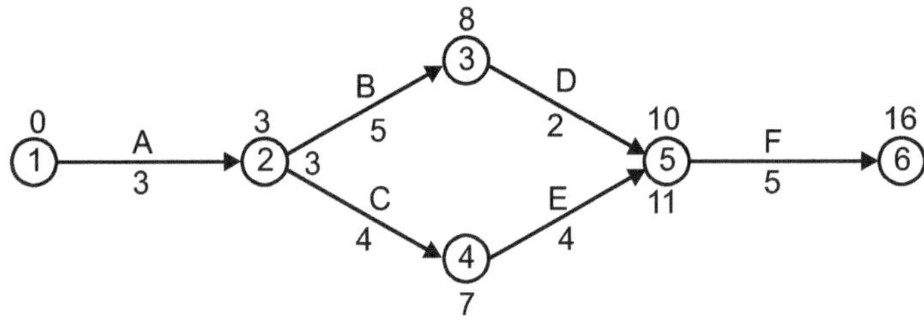

Fig. 2.9

Now, to know LST and LFT of all activities, Backward Pass method to be followed.

Backward Pass :

Activity F is last activity, its EFT 0 = 16. This must be completed at its EFT timing.

∴   Latest Finishing Time (LFT) of activity F = 16. Its duration is 5 days.

∴   It should be started 5 days advance.

∴   Latest Starting Time of activity F = LFT of activity F − T = 16 − 5 = 11.

LFT of preceding activities are the LST of F = 11.

∴   LFT of Activity D = 11.

LFT of Activity E = 11.   LST of D = LFT − T = 11 − 2 = 9.

LST of Activity E = 11 − 4 = 7.

LFT of Activity B = LST of Activity D = 9.

LST of Activity B = LFT − T = 9 − 5 = 4.

LFT of Activity C = LST of Activity E = 7.

LST of Activity C = LFT − T = 7 − 4 = 3.

Now, LFT of A. Here two activities B and C are there. So LST of B and LST of C will have to be considered and minimum of the two will be LFT of A.

$$\therefore \quad \left.\begin{array}{l} \text{LST of activity B} = 4 \\ \text{LST of activity C} = 3 \end{array}\right\} \text{ Minimum is 3.}$$

∴ LFT of activity A = 3, LST of activity A = LFT − T = 3 − 3 = 0.

Now, write all times of all activities on the network at proper places (Below activity arrow-line).

Now, final network is shown below and Critical path is marked.

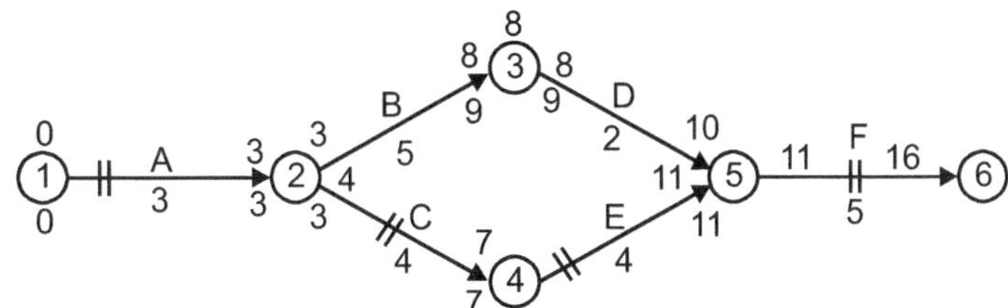

Fig. 2.10

Float Calculations :

Total Float of Activity, A = LST − EST = 0 − 0 = 0

OR   = LFT − EFT = 3 − 3 = 0

Zero Float   Activity is critical.

Activity B :   Total Float = LST − EST = 4 − 3 = 1

OR   = LFT − EFT = 9 − 8 = 01.

                    Total Float = 1 (not zero) ∴ Ordinary Activity.

Activity C :   Total Float = LST − EST = 3 − 3 = 0.

OR   = LFT − EFT = 7 − 7 = 0.

Zero Float. ∴ Activity is critical.

Activity D :   Total Float = LST − EST = 09 − 8 = 1

OR   = LFT − EFT = 11 − 10 = 1.

                    Total Float = 1. ∴ Ordinary Activity.

Activity E :   Total Float = LST − EST = 7 − 7 = 0.

OR   = LFT − EFT = 11 − 11 = 0

                    Total Float = 0. ∴ Critical Activity.

Activity F :   Total Float = LST − EST = 11 − 11 = 0.

OR   = LFT − EFT = 16 − 16 = 0.

Zero Float. ∴ Critical activity.

The critical activities are A, C, E and F. Or 1-2, 2-4, 4-5 and 5-6.

Critical Path is 1 - 2 - 4 - 5 - 6 OR 1-2, 2-4, 4-5, 5-6.

Critical events are (1), (2) , (4), (5) and (6).

Scheduling :

Activity	Arrow	Duration	EST	LST	EFT	LFT	TF	REMARKS
A	1-2	3	0	0	3	3	0	Critical
B	2-3	5	3	4	8	9	1	–
C	2-4	4	3	3	7	7	0	Critical
D	3-5	2	8	9	10	11	1	–
E	4-5	4	7	7	11	11	0	Critical
F	5-6	5	11	11	16	16	0	Critical

Problem 2.4 : *From the following data of a small Civil Engineering Work, Calculate project duration. Draw network and show critical path on it. Calculate EST, LST, EFT, LFT and total float and record in a tabular form.*

Activity	Duration	Activity	Duration
1 - 2 A	5	4 - 8 G	8
1 - 3 B	4	5 - 6 H	6
2- 4 C	3	6 - 9 I	5
2 - 5 D	2	7 - 9 J	2
3 - 4 E	7	8 - 9 K	4
4 - 7 F	4	9 - 10 L	4

Solution : The network is drawn below and the critical path shown on the network. The EST, LST, EFT, LFT and TF are recorded in the table.

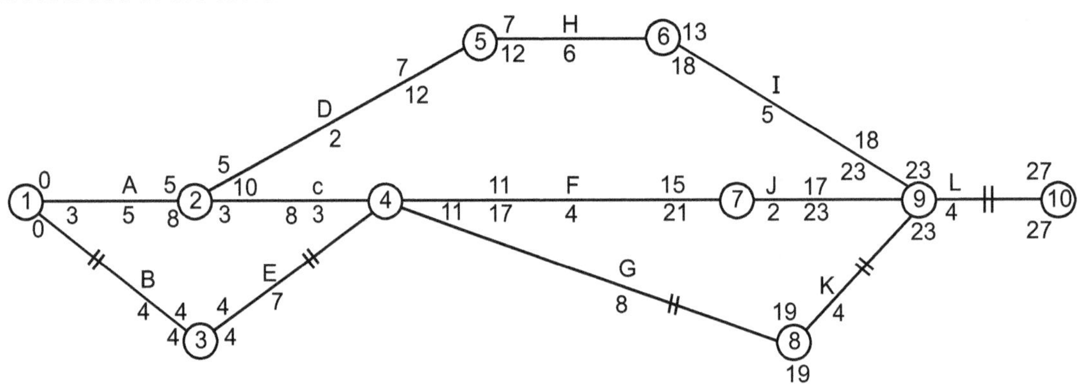

Fig. 2.11

Activity	Duration	EST	LST	EFT	LFT	TF	REMARKS
1 - 2	5	0	3	5	8	3	Ordinary
1 - 3	4	0	0	4	4	–	Critical
2 - 4	3	5	8	8	11	3	Ordinary
2 - 5	2	5	10	7	12	5	Ordinary
3 - 4	7	4	4	11	11	–	Critical
4 - 7	4	11	17	15	21	6	Ordinary
4 - 8	8	11	11	19	19	–	Critical
5 - 6	6	5	12	13	18	5	Ordinary
6 - 9	5	13	18	18	23	5	Ordinary
7 - 9	2	15	21	17	23	6	Ordinary
8 - 9	4	19	19	23	23	–	Critical
9 - 10	4	23	23	23	27	27	Critical

(i) Project duration : 27 days.

(ii) Critical path : 1 - 3, 3 - 4, 4 - 8, 8 - 9, 9 - 10.

Problem 2.5 : *Work out the project duration and establish the critical path for the following network.*

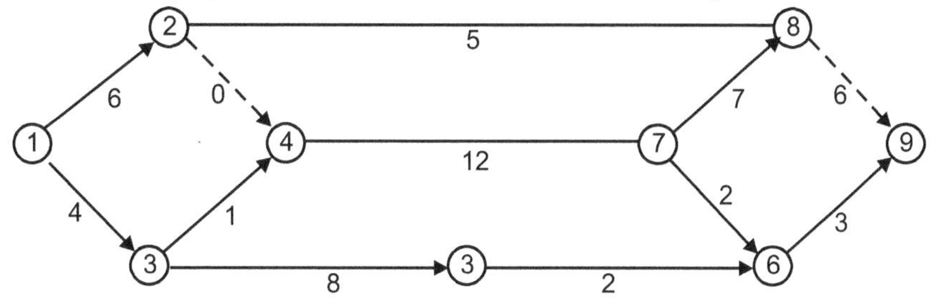

Fig. 2.12

Solution : The critical path is established and shown in the network below.

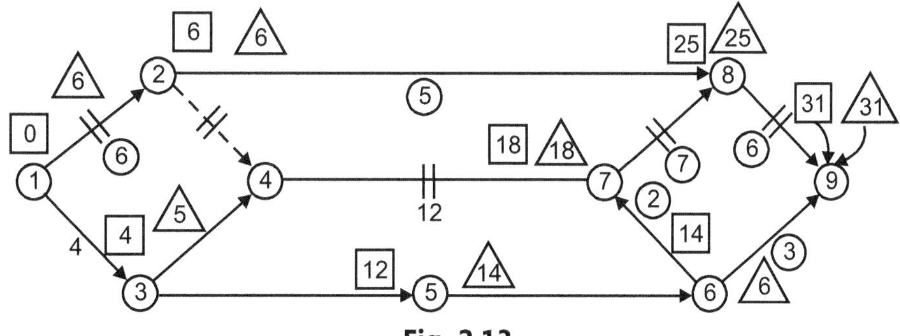

Fig. 2.13

Note :

EST are shown in rectangular block.

LFT are shown in triangles and the critical path represented by conventions in Fig. 2.13.

The project duration is 31 days.

Problem 2.6 : *Given the following network data, draw the network diagram; show the critical path. Also work the EST, EFT, LST, LFT and float and put the same in a tabular form.*

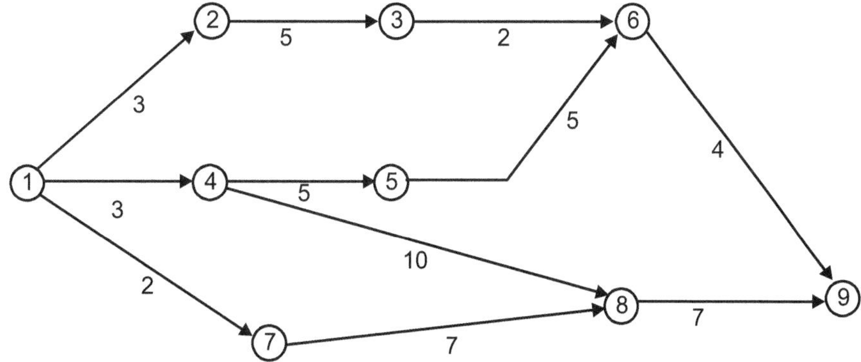

Fig. 2.14

Solution :

Activity	Duration	Activity	Duration	Activity	Duration
1 - 2	3	3 - 6	2	6 - 9	4
1 - 4	3	4 - 5	5	7 - 8	7
1 - 7	2	4 - 8	10	8 - 9	7
2 - 3	5	5 - 6	5		

Note : The critical path is established and shown in the given network.

EST : Early start time;   EFT : Early finish time;   LST : Late start time;

LFT : Late finish time;   EFT : EST + duration;   LST : LFT − duration

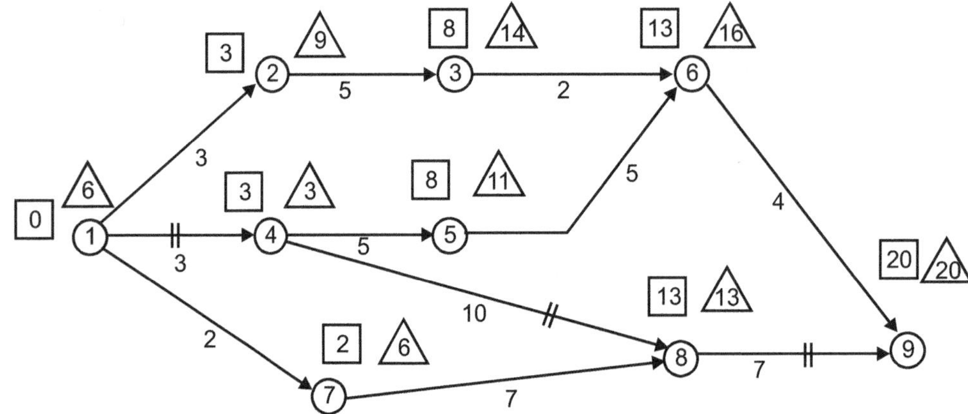

Fig. 2.15 : Network diagram

Activity	Duration	EST	EFT	LST	LFT	Total Float	Remarks
1 - 2	3	0	3	6	9	6	
1 - 4	3	0	3	0	3	0	Critical
1 - 7	2	0	2	4	6	4	
2 - 3	5	3	8	9	14	6	
3 - 6	2	8	10	14	16	3	
4 - 5	5	3	8	6	11	3	
4 - 8	10	3	13	3	13	0	Critical
5 - 6	5	8	13	11	16	3	
6 - 9	4	13	17	16	20	3	
7 - 8	7	2	9	6	13	4	
8 - 9	7	13	20	13	20	0	Critical

Critical Path is 1 - 4 - 8 - 9.

Project duration - 20 days.

Total float is worked out as follows :

$$TF = LFT - EFT \quad or \quad TF = LST - EST$$

If the answer is ZERO the activity lies on the critical path.

Problem 2.7 : *Given the following network diagram. Show the critical path by drawing the network on your answer book. Also work out project duration, EST and LFT.*

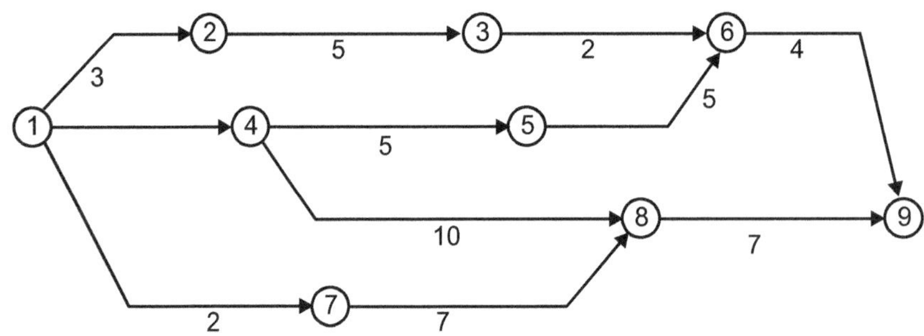

Fig. 2.16

Solution :

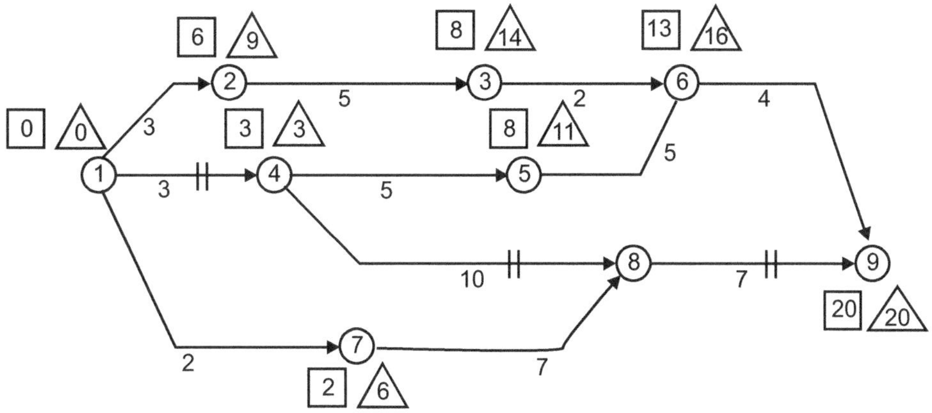

Fig. 2.17

1 - 4 - 8 - 9 is a critical path.

Project Duration : 20 days.

Total float is worked as follows :

$$LFT = EFT \text{ or } LST - EST$$
$$EFT = EST + Duration$$
$$LST = LFT - Duration$$

Activity	Duration	EST	EFT	LST	LFT	Total Float	Remarks
1 - 2	3	0	3	6	9	6	N.C.
1 - 4	3	0	3	0	3	0	Critical
1 - 7	2	0	2	4	6	4	N.C.
2 - 3	5	3	8	9	14	6	N.C.
3 - 6	2	8	10	14	16	6	N.C.
4 - 5	5	3	8	6	11	3	N.C.
4 - 8	10	3	13	3	13	0	Critical
5 - 6	5	8	13	11	16	3	N.C.
6 - 9	4	13	17	16	20	3	N.C.
7 - 8	7	2	9	6	13	14	N.C.
8 - 9	7	13	20	13	20	0	Critical

Note : N.C. : Non-Critical.

Problem 2.8 : *Draw the network from the following data of a Civil Engineering Work. Calculate EST, LST, EFT and LFT and total float for each activity. Mark the critical path and find the project duration.*

Sequence code	Activity	Duration in days
1 - 2	A	9
1 - 3	B	7
1 - 4	C	5
2 - 3	Dummy	0
2 - 5	D	10
3 - 6	E	10
4 - 7	F	12
6 - 8	G	8
6 - 7	I	7
6 - 8	H	15
7 - 8	J	8

Solution : Convention :   EST : Early start time   EFT : Early finish time

LST : Late start time   LFT : Late finish time

Activity	EST	EFT	LST	LFT	Float	Remark
A	0	9	0	9	0	C
B	0	7	2	9	2	N.C.
C	0	5	9	14	9	N.C.
Dummy	9	9	9	9	0	C
D	9	19	16	26	7	N.C.
E	9	19	9	19	0	C
F	5	17	14	26	9	N.C.
G	19	27	26	34	7	N.C.
I	19	26	19	26	0	C
H	19	34	19	34	0	C
J	26	34	26	34	0	C

Note : C → Critical, N.C. → Non-Critical

Critical path is :

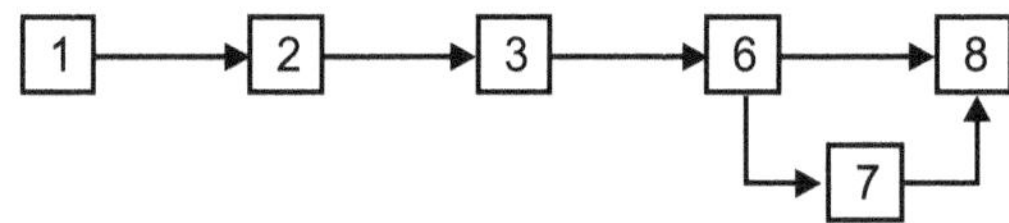

Fig. 2.18

Project duration is 34 days.

Problem 2.9 : *Draw the network from the following data of a small Civil Engineering Work. Calculate EST, LST, EFT, LFT and total float of each activity. Mark the critical path and find the project duration.*

Sequence code	Activity	Duration in days
1 - 2	A	3
1 - 4	B	3
1 - 7	C	2
2 - 3	D	5
3 - 6	H	2
4 - 5	E	5
4 - 8	F	10
5 - 6	I	5
6 - 7	J	4
7 - 8	G	7
8 - 9	K	7

Solution :

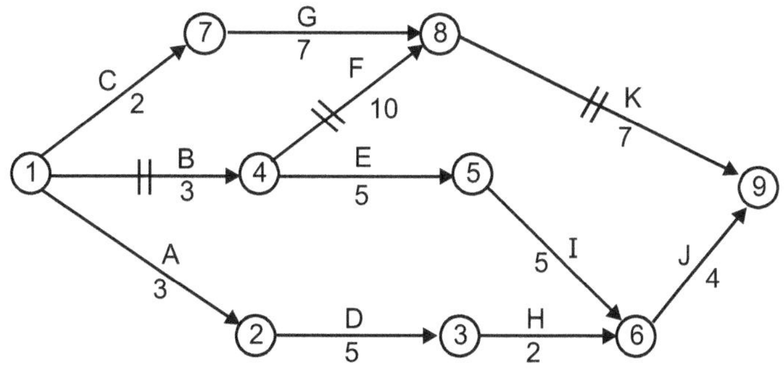

Fig. 2.19

$$\text{LFT} = \text{LST of head event}$$

and

$$\text{LST} = \text{LFT} - \text{T}$$

Total float = LFT − EFT or LST − EST and when total float is zero, the path is critical path.

Sequence	Activity	Duration	EST	EFT	LST	LFT	TF	Remark
1 - 2	A	3	0	03	6	9	6	...
1 - 4	B	3	0	03	0	3	0	Critical
1 - 7	C	2	0	03	4	6	4	...
2 - 3	D	5	3	08	9	14	6	...
3 - 6	H	2	8	10	14	16	6	...
4 - 5	E	5	3	08	6	11	3	Critical
4 - 8	F	10	3	13	3	13	0	...
5 - 6	I	5	8	13	11	16	3	...
6 - 9	J	4	13	17	16	20	3	...
7 - 8	G	7	2	09	6	13	4	Critical
8 - 9	K	7	13	20	13	20	0	–

Critical path : 1-4-8-9.

Problem 2.10 : *Draw the bar chart for "Finalisation of drawings and work order for a building project".*

Activity	Description	Time of completion in weeks
A	Site selection and survey	4
B	Design	6
C	Preparation of drawings	3
D	Preparation of specifications and tender documents	2
E	Tendering	4
F	Selection of contractor	1
G	Award of worker	1

Use your judgement for interdependency of the activities and estimate the project duration.

Solution :

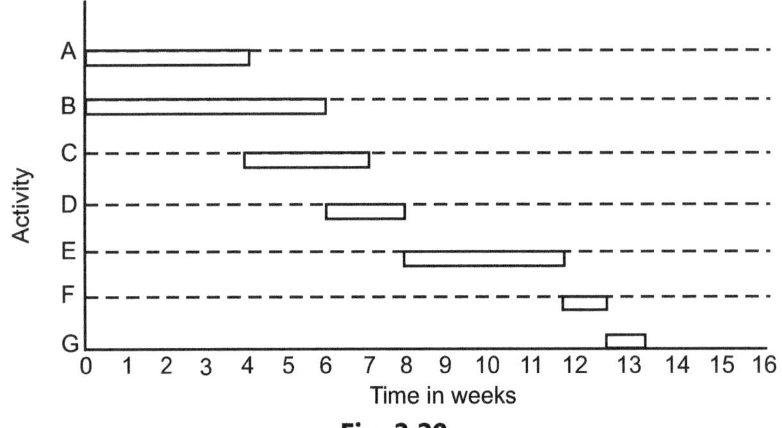

Fig. 2.20

Problem 2.11 : *Draw the Gantt chart for a simple project involving following activities :*
Activity A and B can start simultaneously and proceed parallel.
Activity A is to be completed in 25 days and B is to be completed in 20 days.
Activity C starts 5 days after the starting of activity B and takes 15 days.
Activity D starts at the end of activity C and takes 10 days.
(i)   What is the total project duration ?
(ii) If the progress of activities is as scheduled, show the progress at the end of 15th day on the chart.

Solution :

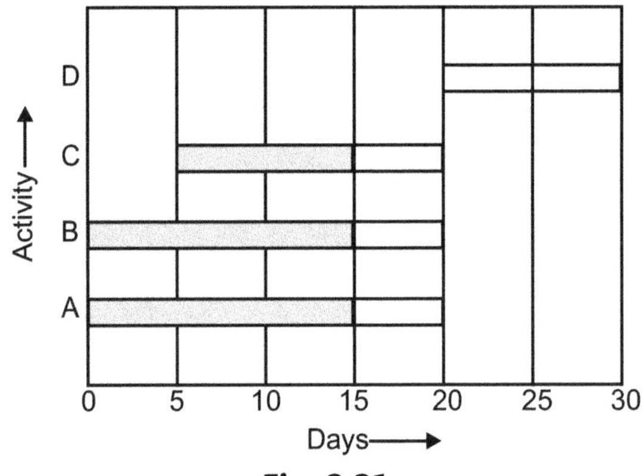

Fig. 2.21

Problem 2.12 : *Draw the network from the following data of a small Civil Engineering Work. Calculate EST, LST, EFT, LFT and total float for each activity. Mark the critical path and find the project duration.*

Solution :

Sequence code	Activity	Duration in days
1 - 2	A	5
1 - 3	B	4
2 - 4	C	3
2 - 5	D	2
3 - 4	E	7
4 - 7	F	3
4 - 8	G	8
5 - 6	H	6
6 - 9	I	5
7 - 9	J	3
8 - 9	K	4
9 - 10	L	6

$$\text{EFT} = \text{EST} + T$$

$$\text{LFT} = \text{LST of head event}$$

$$\text{LST} = \text{LFT} - T$$

$$\therefore \qquad \text{Total float} = \text{LFT} - \text{EFT} \quad \text{or} \quad \text{LST} - \text{EST}$$

When total float is zero the path is critical path.

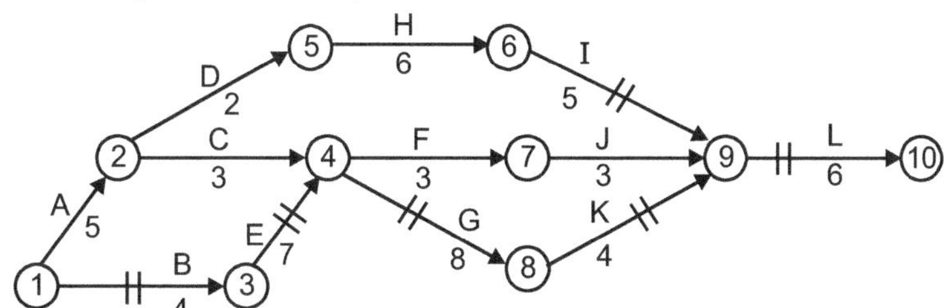

Fig. 2.22

Sequence	Activity	Duration	EST	EFT	LST	LFT	Total float	Remark
1 - 2	A	5	0	05	03	08	03	–
1 - 3	B	4	0	04	0	04	00	Critical
2 - 4	C	3	3	05	08	08	03	–
2 - 5	D	2	2	05	07	10	02	–
3 - 4	E	7	7	04	11	04	00	Critical
4 - 7	F	3	3	11	14	17	06	–
4 - 8	G	8	8	11	19	11	00	Critical
5 - 6	H	6	6	07	13	12	05	–
6 - 9	I	5	5	13	18	18	00	Critical
7 - 9	J	3	3	14	17	20	06	–
8 - 9	K	4	4	19	23	19	00	Critical
9 - 10	L	6	6	23	29	23	00	Critical

Forward Pass :

$$\text{EST}_A = 0 \qquad\qquad \text{EFT}_A = 0 + 5 = 5$$

$$\text{EST}_B = 0 \qquad\qquad \text{EST}_B = 0 + 4 = 4$$

$$\text{EST}_C = 5 \qquad\qquad \text{EFT}_C = 5 + 3 = 8$$

$$\text{EST}_D = 5 \qquad\qquad \text{EFT}_D = 5 + 2 = 7$$

$$\text{EST}_E = 04 \qquad\qquad \text{EFT}_E = 4 + 7 = 11$$

$$\text{EST}_F = 11 \qquad\qquad \text{EFT}_F = 11 + 3 = 14$$

$$\text{EST}_G = 11 \qquad\qquad \text{EFT}_G = 11 + 8 = 19$$

$$\text{EST}_H = 7 \qquad\qquad \text{EFT}_H = 07 + 6 = 13$$

$$\text{EST}_I = 13 \qquad\qquad \text{EFT}_I = 13 + 5 = 18$$

$$\text{EST}_J = 14 \qquad\qquad \text{EFT}_J = 14 + 3 = 17$$

$$\text{EST}_K = 19 \qquad\qquad \text{EFT}_K = 19 + 4 = 23$$

$$\text{EST}_L = 23 \qquad\qquad \text{EFT}_L = 23 + 6 = 29$$

Backward Pass :

$LFT_L = 29$	$LST_L = 29 - 6 = 23$
$LFT_I = 23$	$LST_I = 23 - 5 = 18$
$LFT_J = 23$	$LST_J = 23 - 3 = 20$
$LFT_K = 23$	$LST_K = 23 - 4 = 19$
$LFT_H = 18$	$LST_H = 18 - 6 = 12$
$LFT_F = 20$	$LST_F = 20 - 3 = 17$
$LFT_G = 19$	$LST_G = 19 - 8 = 11$
$LFT_E = 11$	$LST_E = 11 - 7 = 04$
$LFT_C = 11$	$LST_C = 11 - 3 = 08$
$LFT_D = 12$	$LST_D = 12 - 2 = 10$
$LFT_B = 04$	$LST_B = 04 - 04 = 0$
$LFT_A = 08$	$LST_A = 28 - 05 = 03$

Problem 2.13 : *Following are the sequence of activities for a small project. Draw the network diagram and show critical path and find out project period.*

Activity	Dependence	Duration in days
A	Starting activity	2
B	Depends on A	3
C	Depends on A	4
D	Depends on A	5
E	Depends on D	7
F	Depends on B, C and D	6
G	Depends on E, and FG is the last activity	10

Solution : Critical path is denoted by double lines.

Critical path : 1 - 3 - 5 - 6.

Total project duration = 24 days.

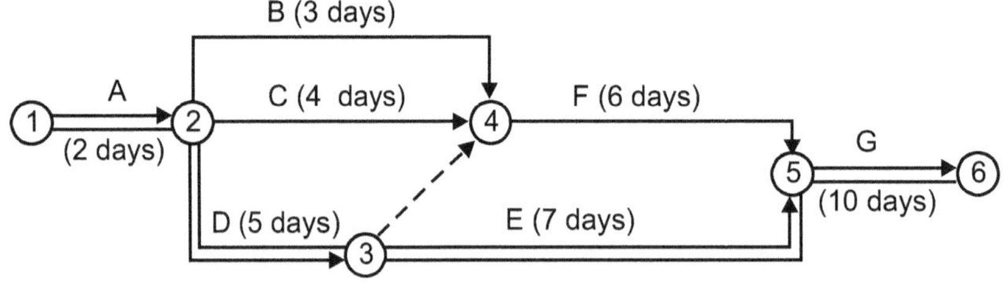

Fig. 2.23

Problem 2.14 : *Draw the network and mark critical path for following activities :*

	Activities	Duration in days
A	1 - 2	3
B	2 - 3	5
C	2 - 4	4
D	3 - 5	2
E	4 - 5	4
F	5 - 6	5

Also calculate EST, LST, EFT, LFT and total float for each activity.

Solution :

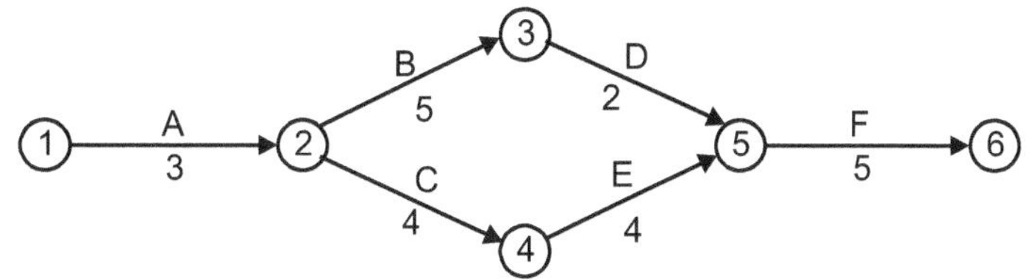

Fig. 2.24

Path	Name of path	Duration (days)
1	A-B-D-F	3 + 5 + 2 + 5 = 15
2	A-C-E-F	3 + 4 + 4 + 5 = 16

Comment : The second path (A-C-E-F) is the critical path because it requires the longest time i.e. 16 days for completion of project.

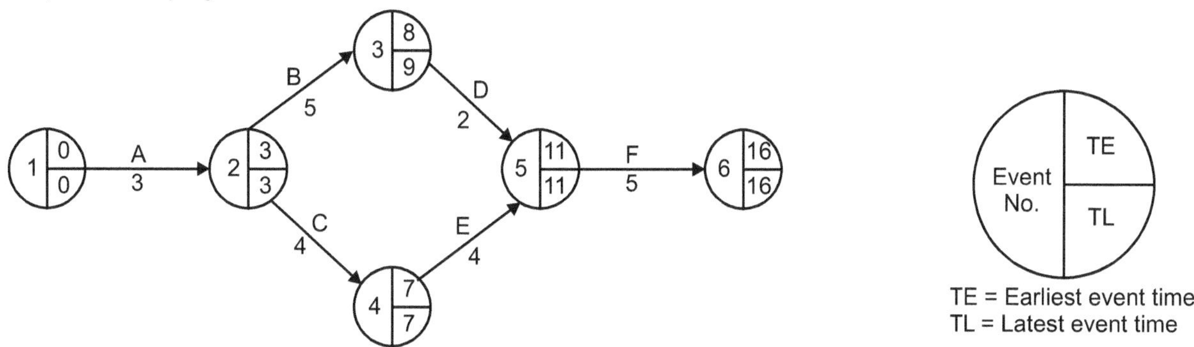

Fig. 2.25

Event No.	Tail event	TE of tail event	Duration of activity	TE of the indicated event
1	none	–	–	0
2	1	0	3	3
3	2	3	5	8
4	3	3	4	7
5	3, 4	8, 7	2, 4	11
6	5	11	5	16

Maximum (8 + 2 = 10, 7 + 4 = 11)

TE of the indicated event = 11.

To calculate TL :

Steps :

1. TL of event 6 = TE since it is end event i.e. 16 days
2. TL of event 5 = 16 − 5 = 11
3. TL of event 4 = 11 − 4 = 7
4. TL of event 3 = 11 − 2 = 9
5. TL of event 2 = Minimum should be taken

    　　　　　　　　 = 9 − 5 = 4

    　　　　　　　　 = 7 − 4 = 3

6. TL of event 1 = 3 − 3 = 0

Activity	Event comprising activity		Activity duration (t_e)	Earliest start (ES_{ij}) = Earliest event time (TE) of tail event	Earliest finish $(EF_{ij}) = (TE_i + t_{ij})$	Latest finish (LF_{ij}) = TL of head event	Latest start $(LS_{ij}) = LF_{ij} - T_{ij}$	Total float (FT) $= LF_{ij} - EF_{ij}$
	Tail (i)	Head (j)						
1 - 2	1	2	3	0	0 + 3 = 3	3	3 − 3 = 0	10
2 - 3	2	3	5	3	3 + 5 = 8	9	9 − 5 = 4	1
2 - 4	2	4	4	3	3 + 4 = 7	7	7 − 4 = 3	0
3 - 5	3	5	2	8	8 + 2 = 10	11	11 − 2 = 9	1
4 - 5	4	5	4	7	7 + 4 = 11	11	11 − 4 = 7	0
5 - 6	5	6	5	11	11 + 5 = 16	16	16 − 5 = 1	0

Problem 2.15 : *Draw the network for the following activities.*

Activity	*Duration (days)*	*Activity*	*Duration (days)*
A (1- 2)	*4*	*D (3 - 5)*	*4*
B (2 - 3)	*4*	*E (4 - 5)*	*3*
C (2 - 4)	*2*	*F (5 - 6)*	*8*

Calculate EST, LST, EFT, LFT and total float for each activity and mark critical path on the network.

Solution :

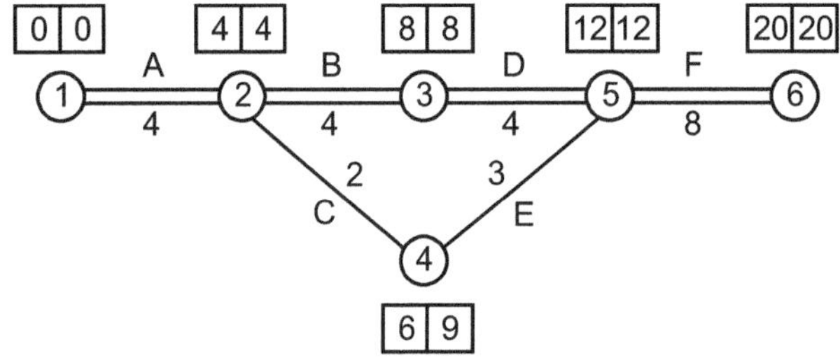

EFT = EST + T

LST = LFT − T

Total float = LFT − EFT

Fig. 2.26

Critical path is 1 - 2, 2 - 3, 3 - 5, 5 - 6.

Sequence	Activity	Duration	EST	EFT	LST	LFT	Total Float
1 - 2	A	4	0	4	0	4	0
2 - 3	B	4	4	8	4	8	0
2 - 4	C	2	4	8	7	9	1
3 - 5	D	4	8	12	8	12	0
4 - 5	E	3	6	9	9	12	3
5 - 6	F	8	12	20	12	20	0

Problem 2.16 : *Draw the network for following activities :*

	Activity	*Duration (days)*
A	*1 - 2*	*4*
B	*2 - 3*	*6*
C	*2 - 4*	*5*
D	*4 - 5*	*7*
E	*3 - 5*	*5*
F	*5 - 6*	*4*

Calculate EST, LST, EFT, LFT and total float for each activity.

Solution :

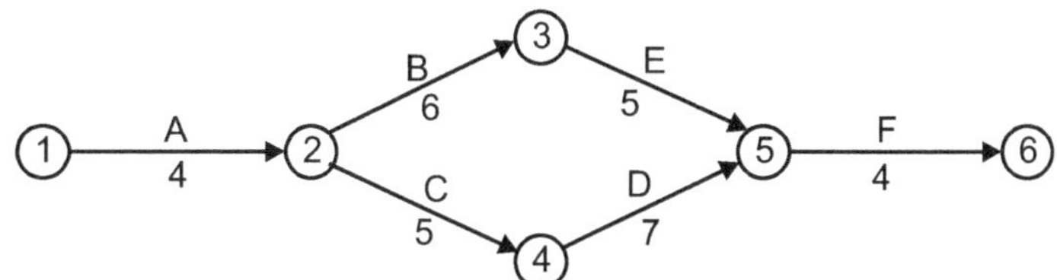

Fig. 2.27

Event No.	Tail event	TE of tail event	Duration of activity emerging event	TE of the indicated event
1	None	–	–	0
2	1	0	4	4
3	2	4	6	10
4	2	4	5	9
5	3, 5	10, 9	5, 7	16 (max. of the two)
6	5	16	4	20

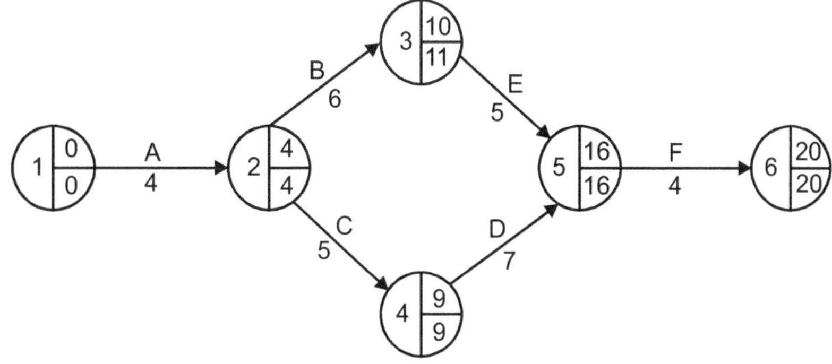

Fig. 2.28

To find out the latest event time (TL) :

1. TL for event 6 = T_E = 20.
2. TL for event 5 = 20 − 4 = 16.
3. TL for event 4 = 16 − 7 = 9.
4. TL for event 3 = 16 − 5 = 11.
5. TL for event 2 = 11 − 6 = 5, 9 − 5 = 4 (Take minimum value).
6. TL for event 1 = 4 − 4 = 0.

Now, to calculate EST, EFT, LST, LFT and total float.

Activity	Event		Event time (TE)	EST	EF	LF	LS	FT
	Tail (i)	Head (j)						
1 - 2	1	2	4	0	4	4	0	0
2 - 3	2	3	6	4	10	11	5	1
2 - 4	2	4	5	4	9	9	4	0
3 - 5	3	5	5	10	15	16	9	1
4 - 5	4	5	7	9	16	16	9	0
5 - 6	5	6	4	16	20	20	16	0

EST = Earliest start = Earliest event time − (TE) of tail event

EF = Earliest finish = EST + TE

LF = Latest finish = Latest event time − TL of head event

LS = Latest start = LF − TE

FT = Total float = LF − EF

Problem 2.17 : *Workout the project duration, EST, LFT and establish critical path for the following network diagram. Draw the network diagram in your answer book.*

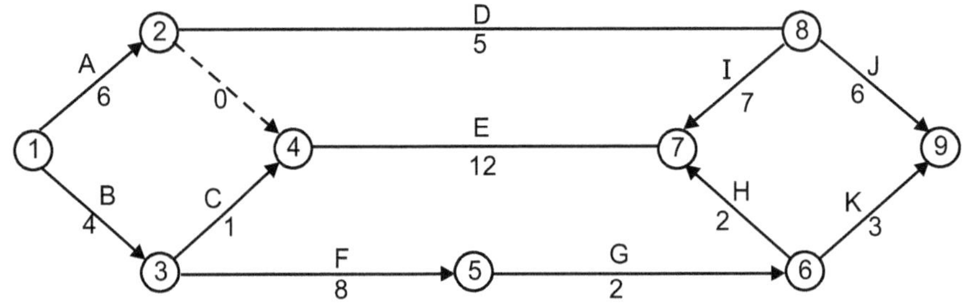

Fig. 2.29

Solution :

No.	Name of Path	Duration
1.	A – D – J	6 + 5 + 6 = 17
2.	A – E – I – J	6 + 12 + 7 + 6 = 31*
3.	A – E – H – K	6 + 12 + 2 + 3 = 23
4.	B – F – G – K	4 + 8 + 2 + 3 = 17
5.	B – F – G – H – I – J	4 + 8 + 2 + 2 + 7 + 6 = 29
6.	B – C – E – I – J	4 + 1 + 12 + 7 + 6 = 30

Path A – E – I – J = 31 is critical.

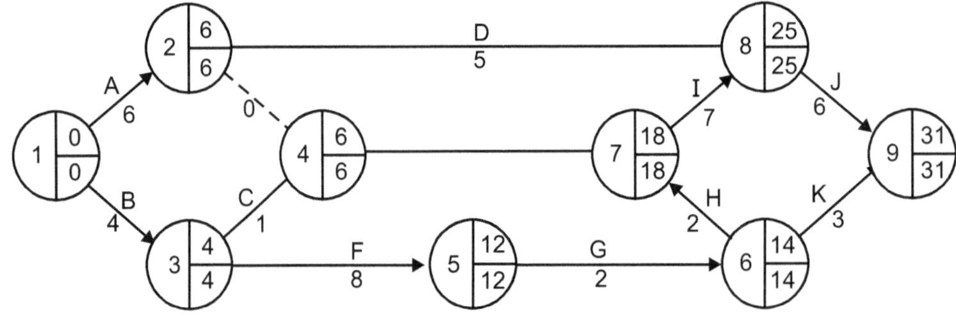

Fig. 2.30

Event No	Tail event	TE of tail event	Duration of activity emerging event	TE of the indicated event
1.	None	–	–	0
2.	1	0	6	6
3.	3	1	4	4
4.	2, 3	4, 6	0, 1	5, 6*
5.	3	4	8	12
6.	5	12	2	14
7.	4, 6	6, 4	12, 2	18, 16*
8.	2, 7	6, 18	5, 7	11, 25*
9.	9, 6	25, 14	6, 3	31, 17*

* Indicates, Take maximum of the two values. Now, to find out TL (latest event time) step.

(a) TL for event $9 = TE = 31$

(b) TL for event $8 = 31 - 6 = 25$

(c) TL for event $7 = 25 - 7 = 18$

(d) TL for event $6 = \begin{rcases} 18 - 2 = 16 \\ 31 - 3 = 28 \end{rcases}$ ** Take minimum of two

(e) TL for event $5 = 14 - 2 = 12$

(f) TL for event $4 = 18 - 12 = 6$

(g) TL for event $3 = \begin{rcases} 6 - 1 = 5 \\ 12 - 8 = 4 \end{rcases}$ ** Take minimum of two

(h) TL for event $1 = 6 - 6 = 0$

and $\qquad 4 - 4 = 0$

Activity	Event comprising activity		Activity duration	Earliest start (EST) = Earliest event time (EE) of tail event	Latest finish (LF) = Latest event time (TL) of head event
	Tail (i)	Head (j)			
1 - 2	1	2	6	0	6
1 - 3	1	3	4	0	4
2 - 4	2	4	0	6	6
2 - 8	2	8	5	6	25
3 - 4	3	4	1	4	6
3 - 5	3	5	8	4	12
4 - 7	4	7	12	6	18
6 - 8	6	8	5	6	25
7 - 8	7	8	7	18	25
6 - 7	6	7	2	14	18
8 - 9	8	9	6	25	31
6 - 9	6	9	3	14	31

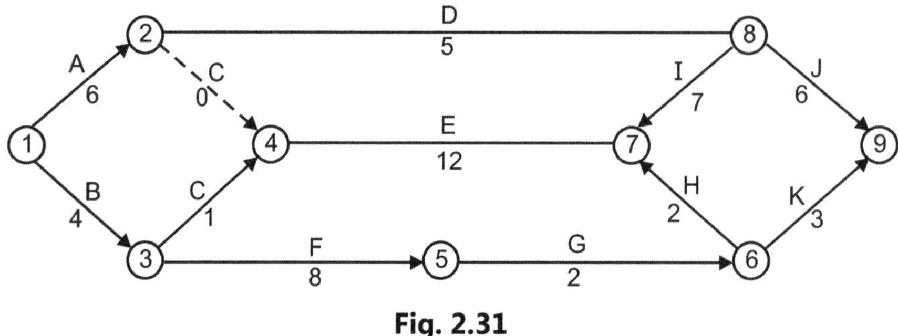

Fig. 2.31

Problem 2.18 : *Draw the network for the following activities and calculate EST, EFT, LST, LFT and mark the critical path.*

Activity	Duration (Days)
A 1- 2	4
B 2 - 3	4
C 2- 4	2
D 3- 5	3
E 4 - 5	4
F 5 - 6	8

Solution :

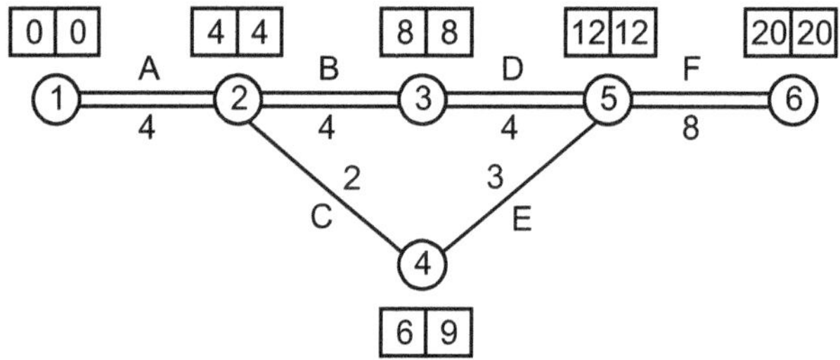

$$EFT = EST + T$$
$$= LST = LFT - T$$
$$\text{Total float} = LST - EFT$$

Fig. 2.32

Activity	Arrow	Duration in days	EST	LST	EFT	LFT	TF	Remark
A	1 - 2	4	0	0	4	4	0	Critical
B	2 - 3	4	4	4	8	8	0	Critical
C	2 - 4	2	4	5	6	7	1	–
D	3 - 5	3	8	8	11	11	0	Critical
E	4 - 5	4	6	7	10	11	1	–
F	5 - 6	8	11	11	19	19	0	Critical

Critical path is shown thus 1 – 2 – 3 – 4 – 5 – 6.

Problem 2.19 : *Draw the network for the following activities and show the critical path.*

	Duration	*Precedes*	*Follows*
A	6	None	B
B	7	A	C, D
C	8	B	E, G
D	5	B	F, I, K
E	6	C	H
F	5	D	H
G	6	C	L
H	8	E, F	J
I	4	D	J
J	6	H, J	L
K	6	D	L
L	5	G, J, K	None

Solution :

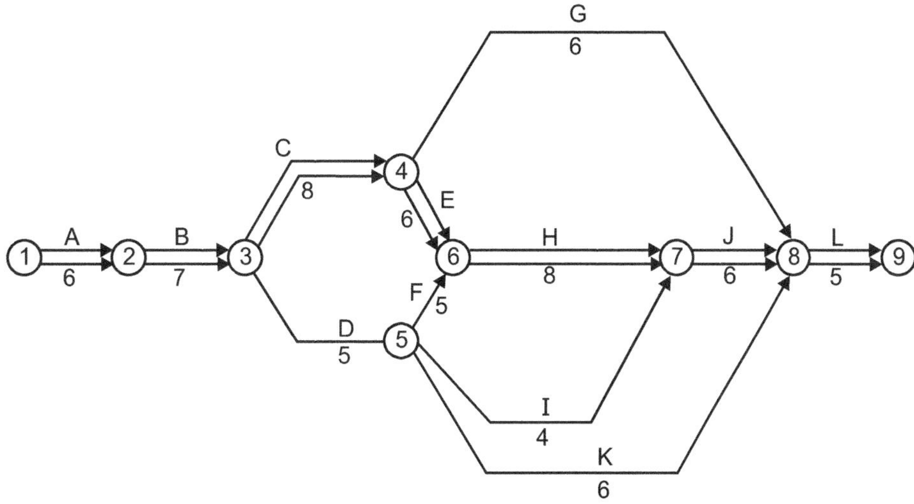

Fig. 2.33

Path		Duration
1 – 2 – 3 – 4 – 8 – 9	→	32
1 – 2 – 3 – 5 – 8 – 9	→	29
1 – 2 – 3 – 5 – 7 – 8 – 9	→	33
1 – 2 – 3 – 5 – 6 – 7 – 8 – 9	→	42
1 – 2 – 3 – 4 – 6 – 7 – 8 – 9	→	46

Critical path is path of maximum duration.

∴   Path 1 – 2 – 3 – 4 – 6 – 7 – 8 – 9 has maximum duration of 46 days.

Hence critical path : A – B – C – E – H – J – L

Problem 2.20 : *Given the network diagram show the critical paths by drawing the network on your answer book.*

1.   *Work out the project duration.*
2.   *Also work out the EST, EFT, LST, LFT and total float.*

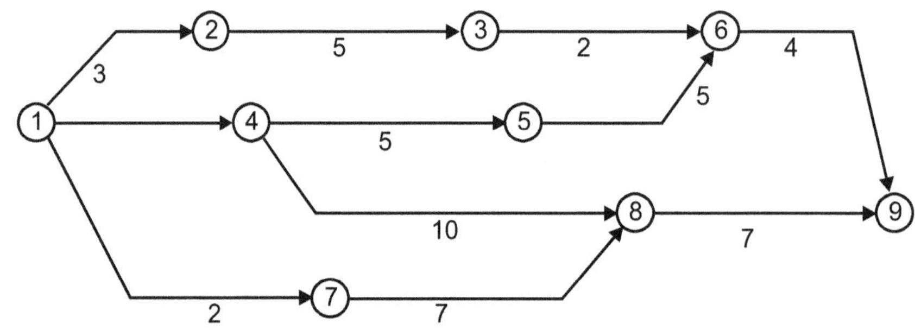

Fig. 2.34

Solution :

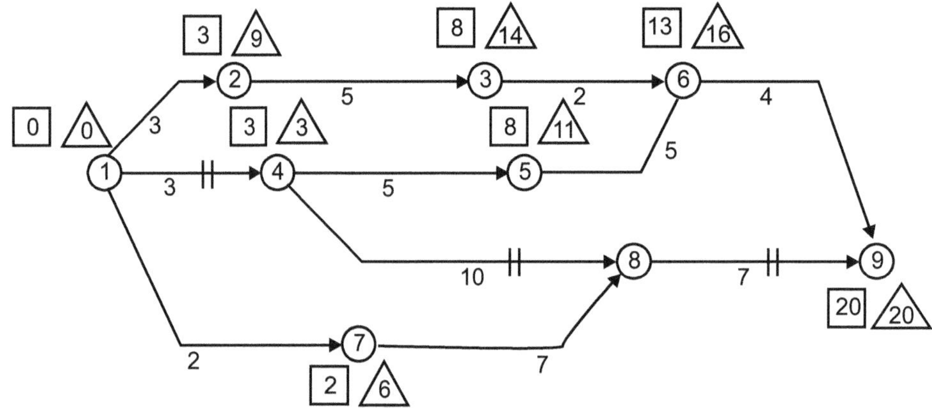

Fig. 2.35

1 – 4 – 8 – 9 critical paths.

Project duration 20 days.

Total float is worked out as follows :

$$LFT - EFT \text{ or } LST \text{ or } LST - EST$$
$$EFT = EST + Duration$$
$$LST = LFT - Duration$$

Activity	Duration	EST	LST	EFT	LFT	TF	Remark
1 - 2	3	0	6	3	9	6	
1 - 4	3	0	0	3	3	0	Critical
1 - 7	2	0	4	2	6	4	
2 - 3	5	3	9	8	14	6	
3 - 6	2	8	14	10	16	6	
4 - 5	5	3	6	8	11	3	
4 - 8	10	3	3	13	13	0	Critical
5 - 6	5	8	11	13	16	3	
6 - 9	4	13	16	17	20	3	
7 - 8	7	2	6	9	13	14	
8 - 9	7	13	13	20	20	0	Critical

2.7 CRASHING OF NETWORK

- Time is the main aspect of network.

- To get quicker benefits of the project, time is the main aspect.

- So it is important to give attention on critical path of the network. This attention is required to complete the project as per schedule or to reduce the completion time.

- While planning the construction projects by network technique, lowest particle cost of each activity is required to consider. So rate of progress of the activity is related to the cost of the activity.

- This progress depends on :

    (i) Delivery of the material.

    (ii) Number of workmen available and

    (iii) Number of type of equipment,

    that are available at minimum cost for each activity.

- When the project is constructed under these conditions it is called as normal programme.

- But sometimes it is required to reduce the total time required to complete the project.

    So in such cases, when normal time is reduced, the project is considered as a crash project and network is a crashing of network.

- So in crashing of network following things are considered :

    (i) Increase the rate of material.

    (ii) Increase the number of workmen.

    (iii) Overtime work.

    (iv) Higher wage.

    (v) Increase the number of unit of equipment.

- Due to this, in crash program, there is increase in the cost of project. So it is required to crash those activities which will permit the desired reduction in construction time at the least total increase in cost.

2.7.1 Purpose of Crashing Network

1. To reduce the total time required to construct a project.

2. To select the activities which can save the time at minimum increase in cost.

3. To get the lowest cost solution by compressing the activity, having lowest cost slope.

4. To get optimum project duration.

2.7.2 Project Cost Analysis and Crashing System

Project Cost Analysis :

- So far the problem of representing a project by a network with a view to determine the critical path has been discussed. The main object of critical path is to identify such activities or events which require special attention in completing the project within the minimum possible time.

- There are several pockets where project duration can be cut down reducing the total project time which has a direct impact on the cost consideration.

- Under certain peculiar conditions, cost, may not be that important, but the time is the most important factor under any circumstances. For example, in the development of a new weapon system cost consideration is less important as compared with the time factor.

- Increase or decrease in the total period required for the completion of a project is closely linked with the increase or decrease in the cost of the project.

Crashing System :

- Controlling phase is effective during construction stages of the project. The progress of work is checked with standard schedules prepared earlier. When the progress seems unsatisfactory, crashing of activities are initiated, time estimates are revised and schedule of work is brought upto date.
- The crashing of activities is quite essential if the project is to be completed within the original schedule. During the revision, the time estimate for some of the activities of the project may be reduced with greater resources. The critical path and critical activities are also required to be defined again.

2.7.3 Important Terms in Crashing

1. **Normal Duration or Time :** It is the duration of an activity or a project under normal conditions.
2. **Normal Cost :** It is the cost of an activity or a project under normal conditions.
3. **Crash Duration or Time :** It is the duration of an activity or a project after crashing.
4. **Crash Cost :** It is the cost of an activity or a project after crashing.
5. **Direct Cost :** It is the cost of material, labour and equipment used in the project.
6. **Indirect Cost :** It is the expenditure incurred on administration, interest on loans, supervision etc. for project.
7. **Cost/Time Slope :** It is defined as, "crashing cost per unit of time" and is expressed as follows :

$$\text{Cost/Time Slope} = \frac{(\text{Crashing cost } (C_c) - \text{Normal cost } (C_n))}{(\text{Normal duration } (d_n) - \text{Crashed duration } (d_c))} = \left(\frac{C_c - C_n}{d_n - d_c}\right)$$

A and B are crashed and normal points for the crashing of an activity. The variation between the two points may be a straight line, or parabolic.

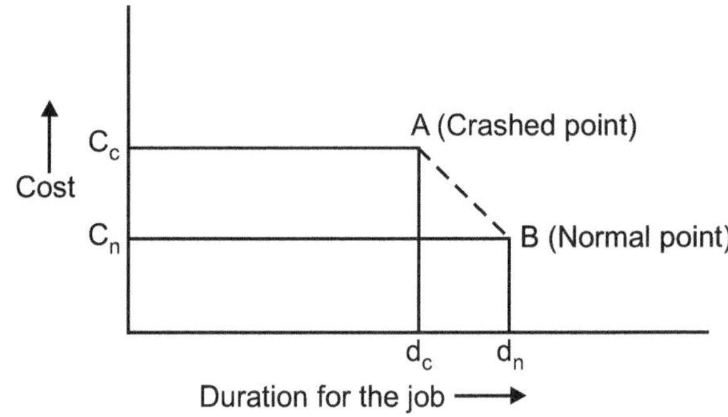

Fig. 2.36

8. **Total cost :** It is the sum total of the direct cost and indirect cost.

9. **Optimum cost :** It shows the variation of total cost of project with project duration. This curve will have point where the tangent would be horizontal. At this point, the cost of the project may be minimum. This minimum cost is called as *Optimum cost* of the project.

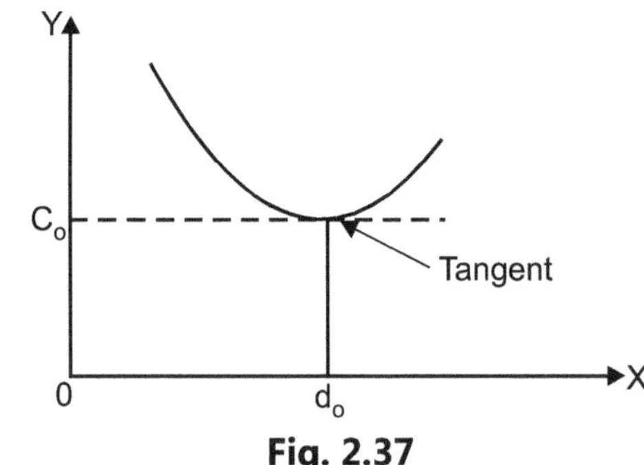

Fig. 2.37

10. **Optimum Duration :** The optimum duration of the project means the time corresponding to optimum cost. It is shown by the point d_o in the above figure. It is necessary to achieve the optimum duration while decreasing the time of the project.

2.7.4 Updating of Schedules in C.P.M. Network Analysis

- In many cases it is necessary to alter the original arrow diagram as the work progress. Under most of the industrial situations, many unknown factors come up when things appear to be going according to plan.
- A key supplier fails to deliver important material or a major design fault is discovered which may take time to put it right. A customer may ask for modification in the design when fabrication and manufacturing processes are already advanced.

- If the network is to be kept dynamic, it must accept such unforeseen changes. The schedule must be updated as soon as possible. Managers can consult the schedule planner to see how much the remaining float the activity possesses. Updating provides the project manager with an opportunity to make changes in the net work.

- The process of incorporating the changes and rescheduling or replanning is called as updating.

2.8 MATERIAL MANAGEMENT

2.8.1 Introduction

- Materials Management is a distinct area of management. The subject matter is very comprehensive. It covers a wide field of knowledge. It comprises of the following departments or sections :

1. Material Planning.
2. Purchasing.
3. Store Keeping.
4. Inventory Control.
5. Production Control.
6. Materials Handling, Shipping and Transport.

- The primary function of management is to employ resources very efficiently and increase productivity. For the same investment of resources it should be possible to increase the margin of profit and production. Materials Management reduces material cost, avoids locking up huge capital, improves profit turn-over ratio. It deals with the whole aspect of material cost; supply and utilisation.

2.8.2 Importance of Materials Management

- If one examines the sales value of any product or the construction cost of any work, it is found that the materials which have gone into the product or construction, account for a major portion of the total cost. Therefore, it is imperative that we should give the greatest importance to the materials management. Even small savings in material cost will effect appreciable and increased profit. Particularly, in the field of industry, top priority is to be given to materials management to achieve economy.

- Unfortunately, many industrial establishments have not yet realised the importance and scope of materials management. In the field of civil engineering, its importance has not been realised at all. Even a saving of 5% in material cost will substantially enhance the profit margin. Many organizations do not have a separate department dealing with materials management. But, in most of the American Companies or Organizations, the materials manager enjoys the second rank or place in the organizational set-up. He does a variety of functions. He is sometimes given the designation "Vice President - Materials". He is the key person who can bring about a substantial increase in the profits of the organization, by reducing materials cost without any trouble from labour.

2.8.3 Objectives of Materials Management

- The materials management is a staff function. It promotes and helps in realising the objectives of an organization. The objectives are :

1. Survival and growth of the organization.
2. Service to customers.
3. Good working conditions.
4. To provide materials of required quality and quantity as and when required at the minimum possible cost.
5. To reduce investment on inventories.
6. To reduce the cost of production or cost of construction, by practising standardisation value analysis, substitution by locally available materials and so on.
7. To reduce unnecessary paper work.
8. To promote skill and knowledge of the persons working in the organization - development of personnel.
9. To function as the watch dog of the organization.

2.8.4 Functions of Materials Manager

- Following are the functions of materials manager :
  1. Materials planning and Programming.
  2. Purchasing.
  3. Store keeping.
  4. Inventory control.
  5. Receiving and Warehousing.
  6. Value analysis.
  7. Transportation.
  8. Material handling.

2.8.5 Equipment of Materials of Construction and Phasing of the same in Relation to the programme of Construction

- We have to prepare various schedules such as labour schedule, equipment schedule, material schedule etc. from the construction programme.

- Before starting the work, detailed quantities of various materials required for the different items of work are worked out from the material schedule and specifications. And accordingly materials are to be kept ready in time for use.

2.9 INVENTORY

- What is Inventory ? It is defined as, "the idle resources of an enterprise or organization". Inventory includes raw materials in process, finished, packing, spare parts. These are held in stock to meet any expected demand or consumption. It is a list of materials, spare parts of machineries giving the quantity and value of each item.

- Inventory is necessary for the following :
  1. To provide good service to customers.
  2. To take advantage of discount by purchasing materials in bulk.
  3. To serve as a buffer stock in the event of delay in obtaining materials.
  4. To allow flexibility in production.
  5. To reduce cost on transport.

2.9.1 Inventory Control

- This means a planned way of finding out what to indent, when to indent, how much to stock. The production schedule should not be disturbed. As mentioned earlier, inventory includes all the raw materials, general stores machinery, spare parts, components purchased or manufactured for stock, work in progress and finished stage.

- Non-availability of inventory when needed leads to what is called a stock out. The stock out is most undesirable. It means that the men and machines in the organization will have no work. The customers cannot be served properly and they will have to wait indefinitely. All these undesirable developments lead to loss of good-will, reputation, and of course, stock-out results in loss and profit too. To avoid all these things, the organization must carry inventory.

- Purchase of materials in small quantities, just to meet a day's or week's requirement is not generally advantageous. Bulk purchase is very often economical. But the materials management has to establish how much to buy, how much to stock, when to buy and replenish.

- In other words, we are bringing inventory control into picture. Many concerns have been ruined because inventory of these, were not properly regulated or controlled.

- The management is quite reasonable in expecting maximum return on the capital invested. Scientific inventory control helps in achieving this.

$$\text{Return on capital} = \frac{\text{Profit}}{\text{Capital used}}$$

- By increasing profit or reducing the capital, it is possible to increase the return in capital employed. But it is not easy to increase the profit as it depends on many factors. The alternative is to reduce the capital employed. The capital comprises of fixed and working capital. Investment on land, machinery, building is fixed capital. Inventories form nearly 80 to 90% of the working capital.

Stock-out Cost :

- The stock-out cost of an item or material is the loss resulting from the stock out of the item or material. It is not in direct proportion to the cost of the item but to the extent of disruption caused to production or construction. For example, if steel is not available for a month, concreting cannot be done for the R.C.C. work. The work has to be stopped for that duration. The resulting loss is not just the cost of the item but much more than that. A critical spare part may cause production held up, if delivery of the item is delayed, the stock-out cost increases.

2.9.2 Advantages of Inventory Control

1. It creates buffer between input and output.
2. It ensures against delays in deliveries.
3. It allows advantage of quantity discount.
4. It allows to maintain progress of work.
5. It ensures against scarcity of materials in the market.
6. It utilizes the benefit or price fluctuations.

2.10 ECONOMIC ORDER QUANTITY (EOQ)

- The economic order quantity is defined as, "that quantity for which the procurement cost and inventory carrying cost are equal". The combined cost is minimum.

- Economic order quantity may be arrived from procurement cost and inventory carrying cost. Procurement cost decreases as the ordered quantity increases.

  EOQ is given by the expression,

$$Q = \sqrt{\frac{2AS}{i}}$$

where,

Q = Quantity per order in ₹

A = Annual requirement of an item in ₹

S = Cost per placement of order.

i = Inventory carrying charges per rupee, per year, expressed in decimal.

SO = Procurement cost

I = Inventory carrying cost

T = Total cost = $S + I$

P = Minimum total cost

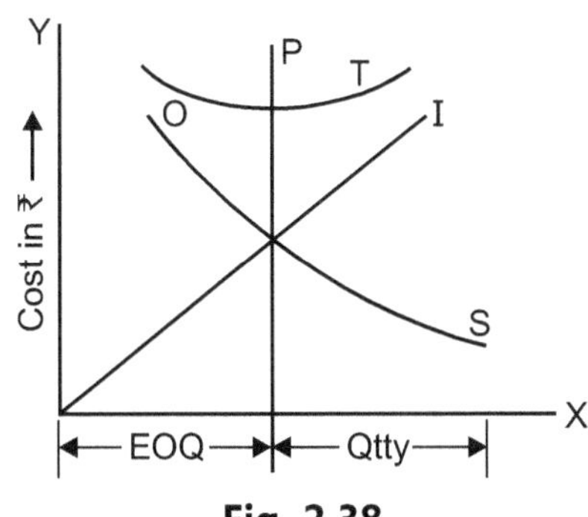

Fig. 2.38

2.10.1 Inventory Carrying Cost

- This consists of expenditure made for storing and handling of material, loss of materials, insurance, loss of interest, depreciation, taxes etc.

2.10.2 Procurement Cost (Ordering Cost)

- This includes salaries and wages of officers in the purchase department, costage, stationary, expenditure for calling quotations, receiving and inspecting materials, payment of bills and other incidental charges.

Example :

- Suppose each consumable article costs ₹ 25/-, 200 numbers are required per year. Cost per placement of order is ₹ 20 and carrying charges is 0.2.

We shall find the quantity per order in rupees.

Here $A = 200 \times 25 = ₹\ 5{,}000, \quad S = ₹\ 20/\text{-} = i = 0.2$

$$Q = \sqrt{\frac{2 \times 5000 \times 20}{0.2}} = ₹\ 1000/\text{-}$$

At ₹ 25/- per piece, number of pieces $= \dfrac{1000}{25} = 40.$

Each time, 40 pieces (or items) to be ordered.

- Consider the following table which shows that the economical order quantity is obtained when the carrying cost and procurement are equal and the total cost is the lowest. In the third column of the table, the average working stock is worked out by taking 50% of each purpose. Annual carrying cost is 20%. Procurement cost @ ₹ 10/- per item.

No. of orders per year	Order size	Average working stock (50%)	Annual carrying cost @ 20% in ₹	Procurement cost @ M 10/-	Total cost
1	2	3	4	5	6
1	3000	1500	300	10	310
2	1500	750	150	20	170
3	1000	500	100	30	130
4	750	375	75	40	115
5	500	250	50	60	110
6	250	125	25	120	145

By developing a table, we can find out the optimum number of orders for different annual requirements of materials.

Example : Find economic order quantity from the following data :

$$\text{Average annual demand} = 30{,}000 \text{ units}$$
$$\text{Inventory carrying cost} = 12\% \text{ of the unit value per year}$$
$$\text{Cost of placing an order} = ₹\ 70$$
$$\text{Cost per unit} = ₹\ 2.$$

Solution : $EOQ = \sqrt{\dfrac{2AS}{i}}$

A = Average annual demand = 30,000 units

S = Cost per placement of order = ₹ 70

i = Inventory varying charges per rupee per year

$$= 2 \times \frac{2}{100}$$

Substituting these values, we get EOQ,

$$EOQ = \sqrt{\frac{2 \times 30000 \times 70}{2 \times \dfrac{2}{100}}} = 4183.3 \text{ i.e. } 4184$$

2.10.3 Bulk Order Quantity

- Bulk order quantity is greater than the economic order quantity. Bulk order quantities are rounded off to 3,6 or 12 month's requirements. But this need not be the same order quantity for all items. For a cheap item it may be one year's requirement; for an expensive item 3 month's requirement may form bulk order quantity.

- Bulk order quantities can be obtained at better purchase prices, the procurement costs are reduced. Work in the purchase department is reduced. But the management will be investing more money. They can be partly compensated by arranging staggering delivery. The staggered rate, of delivery shall not be higher than the economic order quantity.

2.10.4 Arbitrary Order Quantity

- Sometimes, it is not possible to buy bulk order quantity or economic order quantity. It is because of the following situations obtained :

- The market conditions may fluctuate. Prices may steeply rise or fall. It is not always economical to buy E.O.Q. when the price prevailing are high. It is advisable to buy smaller quantities till the prices fall reasonably low. If steep increase in prices is anticipated, it is better to buy a large quantity in advance.

- Sometimes, the rate of consumption may be sporadic. It may suddenly increase to high rate or drop to a low rate. The possibility of placing one order each month may be examined. Some sort of rate contract has to be fixed or the minimum and maximum quantities may be fixed on the basis of the highest probable consumption.

- For items which are of a slow moving nature, it is difficult to fix the order quantity. Perhaps, discretion has to be used in such cases. Six months or twelve months, requirements may have to be obtained each time an order is placed.

2.11 INVENTORY CONTROL : (A-B-C TECHNIQUES)

- As discussed in the previous article the object of inventory control is to keep the investment low and to avoid stock out of critical items. A-B-C technique is a tool to achieve this objective with less cost and effort. It focuses attention on the various items depending on their degree of importance. It enables the management to reduce inventory and to have selective control.

- The inventoryen items are divided into three groups A, B, C depending on their consumption value. Items of high value are grouped under 'A', items of medium value under 'B' and of low value under 'C'. Grouping like this, may vary from enterprise to enterprise and product to product.

   'A' items form 10% in number, 70% of the inventory value.

   'B' items form 20% in number, 20% of the inventory value.,

   'C' items form 70% in number, 10% of the inventory value.

- 'A' items require the greatest control. The minimum stock of 'A' items is kept as far as possible to reduce the capital being tied up, but has to be ordered more frequently. The top executives of the organization must pay attention to the purchase of 'A' items. A constant watch must be kept on the stock position. Minimum stock of 'C' items is kept fairly high liberally (the quantity to last for 6 months or so). No special effort is taken to avoid stock-outs. 'B' items come in between; about 3 to 6 orders are placed per year.

- Suppose there are 1000 items in inventory. 'A' items will be 100 in number, 'B' items 200 and 'C' items 700.

	Value of one month's requirement	Value of three month's requirement	Value of inventory as per A-B-C technique
A items 100	₹ 7,00,000	₹ 21,00,000	₹ 3,50,000
B items 200	₹ 2,00,000	₹ 6,00,000	₹ 3,00,000
C items 700	₹ 1,00,000	₹ 3,00,000	₹ 3,00,000
	Total ₹	30,00,000	9,5000

- In the above table, it is seen that in order to keep three month's requirements of all the three groups of items, the management has to invest ₹ 30,00,000/-. But, if it decides to apply A-B-C technique and keeps $\frac{1}{2}$ month's stock of 'A' items, $1\frac{1}{2}$ month's stock of 'B' items and 3 month's of 'C' items, the value of inventory will work out to ₹ 9,50,000/- (see last column in the above table).

- Thus, instead of requiring ₹ 30,00,000/- on inventory, the management requires only ₹ 9,50,000/- using A-B-C technique. A sum of ₹ 20,50,000/- is reduced. If the bank-rate on overdraft is 12%, an amount of ₹ 2,46,000/- is saved. The capital investment is very much reduced and at the same time profit is increased.

2.11.1 Procurement of Materials

- Materials are procured / obtained either by direct purchase or from store.
- The primary consideration in purchasing is, to procure material at the minimum over-all cost without sacrificing quality. Every organization must have separate purchasing section.
- Purchasing department has the following functions and responsibilities :
    1. To get right quality and quantity of material required for construction.
    2. To make purchase at competitive rates.
    3. To adopt proper methods and follow all government procedures, rules and regulations, see that proper authority/sanction is already obtained.
- Procurement of material means purchasing the materials, construction equipments etc. required for the construction. Procurement depends upon the nature and size of the construction project.

Objectives of Procuring Materials and Stores :
    1. To provide required quantity in right time.
    2. To purchase the economical quantity.
    3. To purchase the materials and equipment from reliable source .
    4. To purchase the materials at reasonable prices.
    5. To receive and deliver materials at right place at right time.

2.11.2 Indent

- Materials such as cement, steel, G.I. sheets etc., or tools required for Government work is to be received on a requisition in the prescribed form (No. 7) called as ' Indent'.
- Materials from stock are demanded in indent form prepared by S.D.O./A.E. incharge of different work as and when required. Indent form is prepared in triplicate, consisting of counterfoil, indent and invoice and is kept in book form serially numbered.
- The counterfoil and indent is filled by the indenting officer and then indent in the invoice are sent to the issuing officer-in-charge of stock who will issue the material as available in stock and shall correct the indent as per actual issue and entries will be made in the invoice and send them to the receiving officer, who will sign the invoice and return it to the issuing officer, as acknowledgement of receipt of which is considered as voucher for the issue.
- In case of issue to the contractor, the cost of materials issued is recovered from the contractor from his hills for the work at the issue rate and at the time of issue, the signature of the contractor should be taken on the invoice. Unstamped receipt of the contractor in triplicate should also be taken giving full particulars of materials, quantities, rates and total cost for recoveries and for enclosing in the bill of cocrator.

2.11.3 Storing

- Storing of materials must necessarily be done systematically so that each type of item has a definite place and can be found when required. A good layout improves the efficiency of the stores.
- Different types of materials are received at the site and are to be stored properly till they are consumed. Any laxity in the care and custody of the store, materials, will increase the unnecessary cost.
    1. Material like sand, stone, metal, bricks etc., can be stored in open area as provided in layout plan.
    2. Materials like cement, fuel, paints, doors etc., are to be stored in sheds.
    3. Perishable materials require careful handling. They are stored in separate sheds.
- In short, it is better to store materials as per the nature and cost of materials.
- Security measures should be taken to prevent theft and pilferage. The common methods are
    1. Providing Enclosure;   2. Double Lock System;
    3. Adopting - Gate Pass System;   4. Providing Security Staff.
- Equally important is the maintenance of store accounts. Proper account is maintained to control over receipt and issue to check shortages, to fix responsibility for the losses etc.
- Stock verification of all store materials and surprise checks are carried out, if required.
- In addition to above, there are other activities such as stock verification, write off losses, road metal account and register and disposal of surplus materials.

2.11.4 Issue of Materials

- All issues are, made against authorised indents. To reduce the routine work issues of different sections are made on different days. The quantities issued should be entered in the register and balance is worked out everytime.

2.11.5 Correct Use and Avoiding Misuse

- Before starting the work, detailed quantities of materials required for the different items of work are worked out from the schedule of quantities and specifications. Materials should be issued and used as per the quantities worked out. Proper account of issue and receipt is kept to control the proper use.

- Misuse of material may be due to careless work, negligence, thefts, want of proper record, control etc. This should be prevented by strict supervision and maintaining proper records.

Methods to Avoid Wastage and Leakage :

- Materials management provides scientific method to avoid wastage and leakage. This can be done in three ways :

    (i) To find out the causes of wastage.

    (ii) To devise suitable norms to prevent wastage.

    (iii) To recover losses on unavoidable wastage.

Causes of Wastages :

(i) Excess purchase.	(ii) Wrong purchase.
(iii) Loss due to faulty storage.	(iv) Loss due to absolescence.
(v) Loss due to rough handling.	(vi) Wastage in consumption.

- Material must be inspected from time to time. Surplus items should be disposed off promptly. On no account deterioration of material shall be allowed to take place. Wastage in consumption is due to carelessness and negligence. This should be checked by proper and strict supervision. Proper storing can help in reducing wastage. Material management helps in purchasing right quantity of materials, carrying out proper studies, and devising proper method, wastage can be checked.

2.12 ORGANIZATION AND ORGANIZING RESOURCES

2.12.1 Introduction

- The activities involved in the construction work require the use and applications of various types of construction equipment. Since earliest time, the builder has sought to develop mechanical devices to facilitate his work in executing engineering works to serve the needs and desires of mankind. From the crude construction equipment utilized by ancient people has evolved the modern construction equipment to seek faster and less expensive methods of construction.

- With the improvement in science and technology and increase an labour rates, there is an increasing trend of mechanization in all the industries all over the world.

- In India, though we have restored to have mechanization particularly on some of our projects, full scale mechanization is still rare.

- Now, it is high time to adopt new construction techniques with new equipment and methods of planning and construction.

2.12.2 Labour Intensive Work as Against Mechanization of Work

- India being agricultural country and having rural population of about 75%, the problem of providing employment to the rural population still fetch the topmost importance. The efforts have been made by implementing the various employment programmes in order to provide more and more employment opportunities to absorb the unemployed youth, however, the problem is not over.

- Considering this background, it becomes necessary to think as to whether we should replace the labour force by mechanization or not ? Mechanization is welcome from the point of view of time factor and cost factor, as it gives speed to the work and reduces the cost of construction. But the problem is to make the choice between employment to the rural people and economy of time and cost.

- The services and utility that will be derived from earlier competition of work may compel to go for mechanization, however, total replacement of raw and skilled labour by automization and mechanization may result in socio-economic injustice. It is, therefore, concluded that labour intensive work may partially be replaced by mechanization which will maintain judicious balance between mechanization and labour intensive work.

2.12.3 Selection of Equipment

For a construction project, the selection of equipment depends upon :

(a) The type, size and other particulars of the equipment.

(b) Whether the equipment is to be purchased, rented or to be procured under hire-cum purchase arrangement.

In both these cases, the final selection depends upon :

(i) Availability of equipment.	(ii) Utility of equipment.
(iii) Production cost of equipment.	(iv) Whether indigenous or imported.
(v) Availability of spare parts.	(vi) Availability of skilled operator.
(vii) Useful life of equipment.	(viii) Duration of project.

2.12.4 Review of Different Types of Construction Equipment and Its Output

- Plants and machinery used in construction industry are given below :

(I) Excavation Equipments :

1. **Tractor :** It is an important equipment for each movement. Tractors are usually worked by diesel engines having horse power ranging from 20 HP to 200 HP.

   They are either :

   (i) Crawler or track type and (ii) Wheel or pneumatic type.

2. **Bull Dozers :** They are used for shallow excavation work and for hauling the earth for short distances of 100 m or so., pushing earth and uprooting trees. There are three types - viz, bull dozer, angle dozer and tree dozer.

3. **Grader :** It is self propelled or towed machine motor grader used for light or medium works. It consists of angled blade 3 to 4 m long supported on a frame work mounted on wheels. It shapes the ground and spreads the loose materials.

4. **Scrapper :** It can dig, load, haul and discharge the material in uniformly thick layers.

5. **Power Shovel :** It is used to excavate earth of all classes except rock, and load it into wagons. They are mounted on crawler tracks. It consists of a mounting, cab, boom, dipper stick, dipper and hoist line. The size of power shovel is indicated by the size of the dipper which is expressed in cubicmetres varying from $\frac{3}{4}$ to $2\frac{1}{2}$ cubic metres.

$$\text{Output of power shovel} = \frac{\text{Volume of excavated material}}{T_e \times N}$$

where,

T_e = Time taken in one cycle of excavation

N = Number of cycles in one hour

6. **Draglines :** They are used to excavate soft earth from below ground and to deposit or to load in wagons.

The output of dragline is measured in cubic metres per hour.

7. **Clam shell :** It consists of a bucket of two halves which are hinged together at top. It is used to excavate soft to medium materials and loose materials.

8. **Hoe :** It can exert high tooth pressures and hence can excavate stiff material which normally cannot be excavated by dragline. Output of hoe is greatest when the excavation is done near the machine, because cycle time of operation reduces.

9. **Dredgers :** They are used for the excavation of the bed of river, lake or sea for the purpose of deepening. Dredgers can excavate upto 50 cu. m/hour and can operate upto 20 m depth with a range upto 35 m.

10. Rippers : They are used for ripping of rock which is an alternative of costly and time consuming drilling and blasting.

11. Trencher or Ditcher : Trenches for mains, gas lines, oil pipe lines, telephone cables, drainage ditches, sewers etc. are dug with the help of trencher machines.

12. Motor Grader : This is used for leveling and finishing earth work.

(II) Compaction Equipment :

* A brief description of each type of compaction equipment is given below :

    1. **Smooth Wheeled Roller :** Three wheeled or macadom rollers and tandem rollers are the typical examples of smooth wheeled rollers. These are most suitable for compacting gravels, sands and such like materials.

    2. **Sheep's Foot Roller :** This type of roller consists of a steel cylindrical drum with steel projection extending in a radial direction outward from the surface of the cylinder. This may be propelled or towed by a tractor. Compaction is obtained by the foot penetrating into the surface. It is more suitable for silty and clayee sand, medium and heavy clay.

    3. **Pneumatic Tyred Roller :** This roller gives kneading action as well as compression to the soil underneath. It is suitable for moderately cohesive silty soils, clayee soils, gravelly and clean sands. The compaction should be done in layers less than 15 cm thickness.

(III) Hauling Equipment :

* Haulage by road is carried out by trucks, rubber tyred tractors with wagons or crawler tractors with wagons.

    1. **Truck :** They have high mobility, good speed and adoptability. The truck capacity generally varies from 0.4 cu.m. to 20 cu.m. and speed may vary from 10 kmph to 100 kmph.

    2. **Dump truck :** These are the trucks which are fitted with automatic unloading devices. The loading is normally done by loading shovels or loaders. The trucks have capacities as high as 53 tonnes. These trucks can be rear dump trucks, bottom dump trucks and side drump trucks.

    3. **Scrappers :** These are important units used also for hauling the material at moderate speed upto 30 kmph.

(IV) Hoisting Equipment :

* These equipments are used for lifting the loads, holding them in suspension during transfer from one place to other and placing them at designated location.

* The hoisting equipments generally used are : (1) Pulleys, (2) Chain hoists, (3) Winches and (4) Cranes.

* Cranes can also be classified as below depending upon the type of mobile units on which they are mounted.

    (i) Hydraulic cranes, (ii) Tower cranes and (iii) Gantry canes.

* Cranes are generally used in steel work, dock yards and railways.

(V) Pumping Equipment :

* Different types of pumps are required depending upon the job requirements. Pump should be dependable in performance. It should be easy to operate and repair. It should be easy to install and economical. Spare parts should be easily available. The different types of pumps are :

    (1) Reciprocating pump, (2) Centrifugal pump, (3) Sludge pump, (4) Submersible turbine pump etc.

* The proper selection of pump will depend upon pumping need at the lowest total cost.

(VI) Equipment used in Concrete Construction :

* Concrete mixing in machine : (1) Continuous mixers and (2) Batch mixer or Drum mixers.

* Continuous mixers are used in massive construction such as dams, bridges etc. Batch mixer or drum mixer is most commonly used and consists of a revolving drum with blades inside it. The batch mixers are of following three types :

    1. Tilting mixers (discharge by tilting) 85 T, 100 T, 140 T, 200 T.

    2. Non-tilting mixers (emptied by chute) 200 NT, 340 NT, 400 NT, 800 NT.

    3. Reversing mixers (emptied by reversin) 200 R, 280 R, 340 R, 400 R.

- The letter T, NT and R used to denote the types of mixer and number such as 85, 100, 200 etc. indicates their nominal mixed batch capacity in litres.

  Vibrators : Types commonly used are :

  (1) Internal vibrators, (2) External or Form vibrators, (3) Surface vibrators and (4) Table vibrators.

  1. Internal vibrators - Used on large works for flat surface.
  2. External or form vibrators - Used for thin sections of walls.
  3. Surface vibrators - Used to finish the concrete surface such as bridge floors, road slabs, station platforms etc.
  4. Table vibrators - Used for consolidation of precast units.

(VII) Crushers :

- Stone crushers and granulator.

(VIII) General :

- Vibrators - Petrol, electric pneumatic, hollow block making machine, wood working machine, mobile crane, mobile building hoist, plastering gun, hollow block cutter, paint spraying equipment, tile making machine, tile polishing machine, sub-mersible pump, welding machine.

(X) For Tar and Bituminous Work :

- The machinery for mixing bitumen and aggregate for applying bituminous materials on road bed and for gritting the road surface afterwards consists of :

1.	Hot mix plant	6.	Bitumen storage tank
2.	Fuel transfer pump	7.	Fuel oil tank
3.	Centrifugal pump	8.	Loader
4.	Dump trucks	9.	Bitumen sprader, Pavar cum fisher
5.	Tandem roller	10.	Light vibratory roller

2.12.5 Owning and Hiring Equipment

- After knowing the types of equipment required for the construction of the project, next most important problem is whether to buy or hire the equipment required for the project.
- If the equipment is to be used frequently and for a long duration of time on the project, contractor has to do cost analysis. The cost of owning and operating the equipment should be carefully analysed and determined so as to decide whether it will be economical or not. The record available of such equipment on similar projects may help as a good guide for this purpose.
- **Following factors are taken into consideration :**
  1. Cost of equipment.
  2. Number of hours it is used per year.
  3. Number of years it is used.
  4. Depreciation.
  5. Cost of maintenance and repairs.
  6. Cost of operation.
  7. Scrap value.
- If it is found that it is economical to purchase then equipment must be purchased.

Hiring the Equipment :

- If the equipment is to be used occasionally and for a short time, it proves to be economical to get it by hiring from firms possessing the same.
- **Hiring the equipment has the following advantages :**
  1. Equipment is available in working condition at short notice.
  2. The investment cost can be used for other better purposes.
  3. No fear to obsolescence of the equipment.
  4. No care is required for the maintenance and storage of the equipment.

2.12.6 Need for Economical Use

- Once the equipment has been acquired it is important that the same is utilized to its full advantages. The idle equipment will add to the losses. Not only that, but it is likely to be rendered unserviceable with the disuse and therefore optimum utilization of equipment is very important.

2.12.7 Maintenance of Equipment

1. The objective of equipment maintenance is to achieve minimum breakdown and to keep the plant in good working condition at the lowest possible cost.
2. Machines should be kept in such a condition which permits them to be used without any interruption.
3. Equipment breakdown leads to an inevitable loss of production.

2.12.8 Record of Equipment :

* Accurate and complete records of equipment is required for many purposes such as cost analysis, control point, accounting purpose and for calculating the cost per unit output. The record is to be kept in standard performa prepared for the purpose.
* It is absolutely necessary to keep records because they are the only reliable guides for measuring the effectiveness of the equipment.
* Record keeping is also necessary for the following :
    1. When budgeting for major overhauling of equipment.
    2. When budgeting for general maintenance cost of equipment.
    3. For finding the reliability of equipment.
    4. For determining schedule of inspection.
    5. For predicting the life of equipment.
    6. For maintenance cost control system.
* Record of equipment should show the following particulars :
    1. Type of equipment and its description.
    2. Name of manufacturer.
    3. Cost and date of purchase of the equipment.
    4. Location of equipment in the project.
    5. Equipment identification, number etc.
    6. Inspection records.
    7. Breakdowns, their dates and reasons.
    8. Cost of breakdown and other associated implications.

Important Points

* The aims of construction management are to execute the work in efficient manner as per drawing, design and specification within the prescribed time-limit and with the maximum possible economy in expenditure.
* Controlling means checking constantly the progress of work with the planned programme and identifying areas of deficiency, if any, so that remedial steps can be taken.
* Scheduling is a graphical device which shows series of jobs to be performed with dates of starting and completing each job or operation as well as sequential relationship among the various jobs.
* Grantt bar chart is the graphical representation of various activities of the work. In Gantt chart, time in days or weeks is marked along the horizontal axis and the activities are represented along the vertical axis.
* **Network diagram :** C.P.M. is a graph of operations. Each operation is represented by an arrow. Every arrow has a tail and a head. The tail of an arrow means the beginning of an operation and the head of an arrow marks the end of an operation.
* Bar charts are inadequate for planning large complex works, where high degree of control is necessary. In such works critical path method is used. When it is very necessary to adhere to the target dates of completion of work critical path method is generally used.
* The C.P.M. is developed by Morgan Walker of Du Pant, US.A. in 1957.
* Time is the main aspect of network.
* To get quicker benefits of the project, time is the main aspect.

- Materials Management is a distinct area of management. The subject matter is very comprehensive. It covers a wide field of knowledge.
- Inventory is defined as, "the idle resources of an enterprise or organization". Inventory includes raw materials in process, finished, packing, spare parts. These are held in stock to meet any expected demand or consumption.
- **Economic Order Quantity (EOQ)** is defined as, "that quantity for which the procurement cost and inventory carrying cost are equal". The combined cost is minimum.
- **Inventory Control (A-B-C Techniques) :** The object of inventory control is to keep the investment low and to avoid stock out of critical items. A-B-C technique is a tool to achieve this objective with less cost and effort. It focuses attention on the various items depending on their degree of importance. It enables the management to reduce inventory and to have selective control.
- Materials such as cement, steel, G.I. sheets etc., or tools required for Government work is to be received on a requisition in the prescribed form (No. 7) called as ' Indent'.

Practice Questions

1. State the necessity and importance of planning in Civil Engineering construction projects.
2. State the various stages in planning.
3. Distinguish between pre-tender planning and post-tender planning.
4. Describe the various activities in contract planning.
5. State the various activities performed from the owners' side in construction project.
6. State the important activities performed from the contractors' side in construction project.
7. State the resources required for execution of construction work.
8. State the meaning of organization. Give the necessity of organization.
9. List out the personnel required on constructing : (i) Cement concrete road, (ii) Earthen dam.
10. Give the wages of the following workers in your locality : (i) Mason, (ii) Scaffolder, (iii) Bar-bender, (iv) Bhistee.
11. State two examples of each of the following categories of worker : (i) Unskilled, (ii) Semi-skilled, (iii) Skilled.
12. State the modes of employment of labour.
13. State four welfare activities which are to be provided to the workers to increase the productivity.
14. State the peculiarities of line and staff organization.
15. Give an organizational set-up of a contracting firm undertaking the construction of masonry dam, mention one important function of each personnel in the organization.
16. Give the organization set-up of P.W. and H.D.
17. State the leadership qualities required in a junior engineer while carrying out his duties.
18. State the functions of personnel management.
19. Enumerate the objectives of material management.
20. Give the expression used for economic ordering quantity and write the meaning of letters used in the expression.
21. State the meaning of Inventory and Inventory Control.
22. State the precautions with respect to quality, you would take as an engineer in purchasing the following materials : (i) Cement, (ii) Bricks, (iii) Timber, (iv) Steel.
    Give the layout for stocking the above materials with reference to the work site.
23. State the advantages of A-B-C analysis.
24. If a Civil Engineer is placed incharge of a construction firm state his function as a manager.
25. 10 cu.m. of structural concrete M 150 is to be placed in one shift of eight hours. List the equipment, labour and materials which should be kept ready.
26. State the causes for wastage materials. Give the measures to reduce the wastage.

27. State the action you will take if the Government stores in your charge are found to have been stolen.
28. State the purpose of indent.
29. State the meaning of "Economic Order Quantity". Give reasons for its use in the management of stores.
30. State the meaning of 'Materials Management'.
31. Explain with graphical representation A-B-C. analysis.
32. Given that :
    (i)   Annual usage = 50 units.
    (ii) Procurement cost = ₹ 15 per order.
    (iii) Cost per piece = ₹ 100 per item.
    (iv) percentage including expenditure on obsolescence, taxes, insurance, deterioration etc. = 10%.
    Calculate EOQ.
33. Determine the economic order quantity from the following data :
    Average annual demand = 10,000 units.
    Inventory carrying cost = 20% per rupee value per year.
    Cost of placing an order = ₹ 100.
    Cost per unit = ₹ 5.
34. State with reasons whether you will choose labour intensive work or mechanical work for the following items of work :
    (i)   Earth work for a big earthen dam.
    (ii) Concreting for a slab overing the roof of a room.
    (iii) Tunnelling in hard rock.
    (iv) Masonry work from medium masonry gravity dam.
35. Mention any four steps leading to the economical use of equipment.
36. State the comparative merits and demerits of hiring the equipment over owning the equipment.
37. List out the equipment required for construction of a large earthen dam.
38. State the precautions you will take to have optimum utilization of equipment and machineries.
39. State the factors affecting the selection of equipment.
40. Name any four excavation equipment and state where they are used.
41. Name any four hauling equipment and state where they are used.
42. State the compaction equipment used for compacting : (i) Clayee soils, (ii) Murum soils, (ii) Gravely and sandy soils.
43. Enlist the equipments used in concrete constructions.
44. State four advantages and two shortcoming of Gantt Chart.
45. Define the term critical path.
46. Give four advantages of network analysis over bar-chart method.
47. State meaning of the following terms used in network : (1) Event, (2) Dummy activity, (3) Total float, (4) EST, LST, EFT, LFT.

Unsolved Problems

1. Calculate the following data of a small Civil Engineering Project, calculate the project duration. Draw network and show critical path on it. Calculate EST, LST, EFT, LFT and total float. Record it in tabular form.

Activity	Duration	Activity	Duration
1 - 2	5	4 - 8	8
1 - 3	3	5 - 6	6
2 - 4	3	6 - 9	5
2 - 5	2	7 - 9	2
3 - 4	7	8 - 9	4
4 - 7	4	9 - 10	4

2.

Activity	Commences after	Duration in weeks	No. of labour required
A	Starting activity	2	3
B	Starting activity	2	6
C	Starting activity	1	4
D	2 weeks	4	2
E	6 weeks	1	4
F	2 weeks	5	1
G	2 weeks	8	4
H	7 weeks	4	2
I	4 weeks	3	2
J	1 week	3	5
K	10 weeks	5	2

Draw bar chart for the project.

3. Draw network using the following data :

Sr. No.	Activity	Duration in weeks	Sr. No.	Activity	Duration in weeks
1	1 - 2	3	6	3 - 4	Nil
2	1 - 3	10	7	3 - 6	3
3	1 - 5	6	8	4 - 6	6
4	2 - 4	6	9	5 - 7	2
5	2 - 5	5	10	6 - 7	4

Carryout forward pass and backward pass, show the critical path in the network.

4. (i) Draw the network for the following data of a small civil engineer project.

Activity	Duration	Activity	Duration
1 - 2 (A)	5	4 - 8 (G)	8
1 - 3 (B)	4	5 - 7 (H)	6
2 - 4 (C)	3	6 - 9 (I)	5
2 - 5 (D)	2	7 - 9 (J)	3
3 - 4 (E)	7	8 - 9 (K)	4
4 - 7 (F)	3	9 - 10 (L)	6

(ii) Calculate EST, LST, EFT, LFT and total float.

(iii) Mark critical path and mention project duration.

5. (i) Given the following network diagram, show the critical path by drawing the network on your answer book.

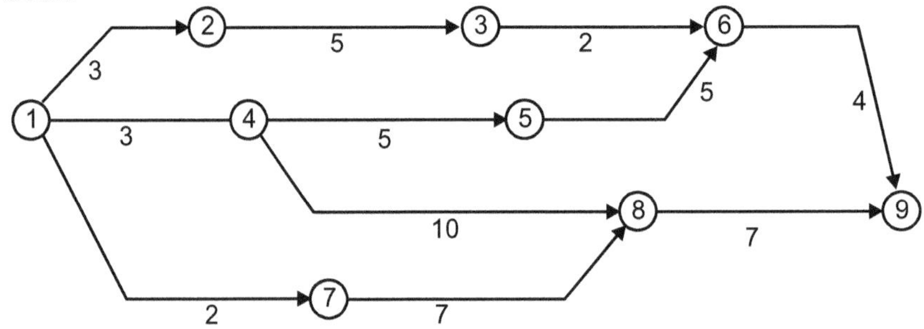

Fig. 2.39 : Network diagram

(ii) Work out project duration.

(iii) Also work out the EST, EFT, LST, LFT and total float.

Chapter **3**

SAFETY IN CONSTRUCTION

3.1   Safety in construction industry - Causes of accidents, Remedial and Preventive measures.

3.2   Labour laws related to construction industry.

Objectives

After learning this chapter, student will be able to

- Identify causes of accidents at construction site in the given situation with justification.
- Suggest safety measures to avoid accidents for the given construction site.
- Apply relevant labour law/s in the given situation of a construction industry.

3.1 SAFETY IN CIVIL ENGINEERING

Introduction :

Safety Engineering "is a term for all efforts on the part of the management to prevent accidents. Safety in construction is considered important for the following reasons :

Importance of Safety :

1. Human life is considered invaluable.
2. Direct costs of previous accidents :
    - Medical care and compensation to the injured.
    - Expenditure on Insurance premium.
    - Cost of legal proceedings.
    - Damage to property.
    - Delay to the project.
3. Indirect costs :
    - Investigation into cause of accidents.
    - Loss of skilled men/women.
    - Loss of equipment.

Definition of Safety :

- It involves a three pronged requirements.
    1. Freedom from danger or risk.
    2. Freedom from injury.
    3. Harmlessness which is the state of being without hurt, loss or liability.
- In construction work, it is essential that all care is taken to ensure safety of all concerned. Preventive measures should be taken to eliminate or minimise accidents resulting in danger to life and property of people concerned.
- Accident is defined as, "an unintentional or unexpected happening except in the case of acts of Nature and God". Accidents are caused by negligence.

- While an accident might occur and endanger the life of a person or lives of persons, the negligence might be his/their own or due to the negligence of others. However, accident prevention regulations are the responsibility of the management, the Contractor or the Engineer in charge or both.

- Factors to judge the nature and extent of accidents in construction work.

- Construction manager will be responsible for all reasonable protection to prevent damage, injury or loss to :

    o   All employees on the work and all other persons who may be affected thereby.

    o   All the work and all material and equipment related for the construction and under the case, custody and control of the construction manager i.e., the engineering in-charge in case of the departmental agency executing the work or the contractor if the work is executed through contract.

    o   Other property at the site or adjacent thereto.

3.1.1 Common Causes Leading to Accidents on Construction Sites and Precautions to be Taken to Avoid Accidents

Causes of Accidents	Safety Measures
Personal Lacking : 1. Lack of ability 2. Lack of knowledge 3. Impairment of functions 4. Fatigue 5. Lack of interest 6. Deliberate taking of risk	Arrange meeting from time to time to discuss safety measures. Display effective safety posters. Maintain good relation in the organization. Arrange films on safety precaution. Provide safety training programme to workers. Give process training of complicated construction. Avoid fatigue.
Faulty Working Methods : 1. Striking against objects	1. Provision of sufficient working space. 2. Removal of any obstructions in the working place.
2. Falling of objects	1. Careful handling of tools and materials. 2. Use of helmets while working at higher levels.
3. Falling of a person	1. Provision of guard rails while working at higher levels. 2. Proper construction of working platforms.
4. Failures of forms, ladders, scaffolds etc. (This is the common cause)	1. Use of strong material. 2. Proper methods for erection especially in high scaffolds. 3. Rechecking of the strength from time to time.
5. Hazards in use of explosives.	1. Careful storage, handling and use of explosives. 2. Vacation of explosion area. 3. Counting the number of explosions to check any misfires. 4. The misfires should be dealt carefully.
6. Presence of injurious gases, toxic dust etc. (This is likely in tunneling operations).	1. Use of gas masks. 2. Proper ventilation. 3. Medical aid.
7. Unforeseen causes.	1. Rescue measures and devices.

Structural Failure :	
1. Failure of sub-stratum.	1. Careful sub-surface investigation. 2. Providing suitable type of foundations. 3. Preventing excavations in the loaded substratum.
2. Failure of excavations.	1. Careful stability investigation. 2. Providing slope or support to the sides. 3. Quick completion of foundation and side refilling.
3. Substandard material.	1. Testing of materials (Testing of cement, concrete cubes, bricks, steel bars, road metal etc.) 2. Sorting out good material (Washing of sand, Segregation of good bricks etc.).
4. Faulty implementation of designs.	1. Inspection by the designer. 2. Competent supervision.
5. Faulty construction practices.	1. Employment of competent trade workers. 2. Competent supervision.
6. Unauthroised changes.	1. Changes to be made only on written orders from competent authority.
7. Carelessness in supervision.	1. Fixing the responsibility of supervision. 2. Written certificates from competent authority at important stages (passing of foundation bed before concreting, passing of reinforcement and form work before concreting.)
8. Natural causes like earthquake.	1. Assessment of the magnitude and suitable provisions in the design of structure.

3.1.2 Important Points, the Construction Engineer and Supervisor has to Bear in Mind in Order to Ensure Safety

1. Understand the drawings and/or specifications. One should not hesitate to ask for clarification in case of doubt.

2. Study the site, think of appropriate modification when the site conditions prove to be different from initial assumption.

3. While undertaking any construction activity, prepare a checklist to include all points you have to check.

4. Select the right workmen for the right job especially when it involves hazardous work.

5. Be sure of the rules which workers should not violate in their own interest. Provide on opportunity to them to understand the rules. No use of having rules written boldly and displayed, when you are dealing with unskilled and illiterate workers.

6. In a sequence of activities, take remedial measures to correct mistakes as and when the activities occur as far as possible.

7. Removal or dismantling of temporary structures is as important as construction of the same.

8. In a demolition work, understand the structure you are intending to demolish. Be certain about the probable consequences of the technique and ensure that precautions are taken.

9. In case an accident does occur, be prepared to provide appropriate aid to the affected workmen in the quickest **possible time**.

3.1.3 Accident

- **Accident** is defined as "an unintentional or unexpected happening". Except in the case of acts of nature and God, accidents are caused by negligence i.e. the failure to exercise such care as would normally be expected of a reasonable man.

  Following factors are sufficient to judge the nature and extent of accidents :

  (1) Injury Frequency Rate. (2) Injury Severity Rate.

- The Injury Frequency Rate is defined as, "the number of disabling injuries per 10,00,000 man hours worked". It is expressed by the equation :

$$\text{I.F. Rate} = \frac{\text{Numer of disabling injuries} \times 1000000}{\text{Number of man hours worked}}$$

- The relationship does not take into account the time lost due to an injury. Injury Severity Rate is defined as "the number of days of lost time because of injuries per 1000 man hours worked". It is expressed by the equation :

$$\text{I.S. Rate} = \frac{\text{Number of days lost} \times 1000}{\text{Number of man hours worked}}$$

- According to some classifications, construction accidents are grouped as follows :
  1. Uncontrolling contact between workers and equipment.
  2. Failure of temporary structures such as scaffolds, ramps, ladders, form work etc.
  3. Inherent engineering hazards e.g. use of explosives in blasting operations.
  4. Unsafe practices of workers, carelessness while working.

Accident Cost :

- Every accident brings with it losses in the form of sacrifice of human lives, loss of material or equipment, injuries to workers, damages to other properties, loss of time and ultimately loss of money. Thus in short, an accident increases the cost of construction and decreases the margin of profit of the contractor. This cost is known as accident cost.

- The accident cost is broadly classified into two types :
  1. Direct costs or Tangible costs.
  2. Indirect costs or Intangible costs.

1. Directs Costs (Tangible Costs) :

- The costs which can be easily evaluated are known as direct costs or tangible costs. Such costs include the costs required to be paid to the workers for temporary or permanent injuries, or for the loss of equipment or for loss of materials. It can be reduced by taking insurance policy.

2. Indirect Costs (Intangible Cost) :

The costs which cannot be easily evaluated are known as Indirect costs.

The costs required for the following items come under indirect costs.

1. Cost due to extension of time required to complete the work.
2. Cost due to spoilage of materials or equipment.
3. Cost of time lost by injured workers.
4. Cost of time lost by other workers who stop work because of an accident.
5. Cost of loss in production because of an accident.
6. Cost of time lost in dealing work related to accident, reporting, inquiry, cause finding, alternate arrangement etc.

3.1.4 Management's Role

- Top management must also be interested in the safety programmes. Their involvement is equally necessary for implementing safety programmes. In fact, the programme has to originate at the top without which it will not be a success.

- The safety programme must be given publicity. For each job the programme must be developed and installed on a competitive basis. The new employees must be in doctrinated.

- A safe worker may not necessarily be a good worker, but it has been proved often that a good worker is a safe worker. Unsafe acts of employees rather than unsafe conditions are responsible for a large number of accidents.

- The site engineer has to take responsibility and suitable steps to the care, use and maintenance of equipment and machinery. On large works, particularly in hydroelectric projects and tunnel constructions, a large number of construction machinery is deployed.

- People must be conversant in the use of various machinery and the machinery must be well designed and must have safety guards. Every effort must be made to promote good house-keeping. They have to think of safety and practise safety.

3.1.5 Safety in Construction Works

Excavation Work :

- For most of the Civil Engineering works, excavation is one of the first job to be performed. While excavating the trenches, if the sides of the excavation collapse on a few men working on it, it may be their last job. They may not escape from the mishap. There is possibility of all types of ground, collapsing under certain conditions.

Causes of Collapse :

1. Failure of the soil : Soil not supporting its own weight.
2. Moisture breaking down the strength of the soil.
3. Failure due to vibration from the movement of vechicles nearby.
4. Failure due to the weight of loads placed near the edge of the excavation.
5. Failure due to changes in the nature of the soil, presence of pockets of running sand etc.
6. Failure due to excavation being carried out near the site of a previous excavation.

Prevention :

- The one simple way of preventing collapses of the trench, is to slope back the excavation to a safe angle and timbering the sides of the trench. The other measures are that loads should not be allowed at or near the edge of the excavation, heavy materials must not be stacked near the edge, movements of vehicles near the excavation must be avoided.

- Timbering must be done before workers enter the trench. Work should never be carried out ahead of timbering. Every excavation and timbering must be inspected daily.

- Apart from the collapse of soil accidents may be caused due to the following :

1. Workers struck by rock or materials such as pipes falling into the excavation.
2. Workers skidding and falling into the trenches.
3. Unsafe means of access.
4. Workers hit by excavating machinery.
5. Vehicles driven into excavations.

- Excavated area must be fenced or covered. Proper lighting or warning lights must be provided. Secured ladders must be provided to get into the trenches wherever required. Adequate stop blocks or timber planks must be provided to stop vehicles. Some of these measures will minimise and prevent the accidents.

3.1.5.1 Safety in Scaffolding

- Quite a few accidents take place due to scaffold and ladder collapse or breakdown. These can be more or less prevented if precautions are taken during construction and use of scaffolding, if ladders are sound and firmly secured.
- A scaffold must be prepared from sufficient and sound material. On no account, one should try to save material consumption. That is false economy. Faulty materials should not be used. They have dangers concealed in them.
- Before a scaffold is used it must be inspected thoroughly. All weakened or damaged parts must be removed. The standards must always rest on base plates. Bricks and other loose materials are not suitable for the bases of uprights.
- Standards or uprights must be close together and stiffened by rise at regular intervals to prevent movement. Diagonal bracing may have to be used. Over-turning can best be prevented by laying the scaffold using transom or cross tubes, passing through holes in the walls and secured to another tube at right angles which bears hard on the inside of the wall.
- The supports for a working platform on a scaffold must be properly spaced to prevent the planks from sagging and to avoid excessive over-hang, guard rails and the boards must be fixed. Accidents can occur if men remove scaffold planks from a working platform for using somewhere else.

Ladders :

- A ladder should be placed at an angle of 75° to the horizontal. The foot is to be tied so as to be driven into the ground or placed on a sand bag or sole plate to prevent the base from going sideways. A ladder should never be placed on a pile of loose bricks, oil drum or dust bin.
- Gang ways and ramps must be wide enough to all men to cross each other and to allow movement of materials. Slope must be gentle in case of ramps. The boards and guard rails must be provided on gang ways and ramps.

Roofing :

- Injuries are caused by falls from eaves or gutter levels. Falls from eaves level can be prevented by providing a barrier on the roof slope at the eaves. The barrier must extend to a sufficient height and be strong enough to stop a man sliding down the roof slope.
- There must be a working platform below the edge of eaves 30 cms below the eaves and projecting 60 cms beyond the edge of the roof. Roofing material must not be fixed in windy weather. Warning notice must be fixed and must be clearly visible. Accident can be prevented from the knowledge of the earlier happenings.

Safety Programme :

- With the introduction of an effective safety programme, it is possible to avoid most of the accidents. But there cannot be a single programme that can be applied to the construction industry as a whole. It has to be integrated with the various operations of a construction company.
- As the nature and characteristic of C.E. work is unique, every situation needs unique safety programme. It is necessary to chalkout suitable suitable safety programme as per the nature of work. It assures considerable reduction in the injury frequency rate and injury severity rate.
- It may be :
  1. Training programme to workers.
  2. Removing the workers who are frequently making accidents.
  3. Training to some workers to give first aid treatment.
  4. Keeping working place clean.
  5. Physical - medical examination. Instruction programme and training to new workers.
- The programmes are no longer dispensable luxury nor a matter of charity. There is great need for each one of construction industry to try to make it a safe industry to avoid damage and tragedy which results from accidents.
- A safe worker may not necessarily be a good worker but a good worker is always a safe worker.

3.2 IMPORTANT ACTS AND LAWS

Introduction :

* The condition of the labour working in construction industry is very poor. The number of workers employed in construction industry is nearly four crores. They are mostly unskilled or semiskilled workers.

* In construction industry, the contractor dictates the terms of service conditions particularly to skilled workers. They are paid very poor wages. They have very little or practically no bargaining power. It leads to their exploitation.

* They are migrating from one place to another for getting work. They remain illiterate. Under these circumstances labour force in construction industry cannot get organised as in the case of other factory workers.

* Due to poor working conditions, their efficiency also remains below the expectancy. To remedy the situation, the Government has, from time to time, brought out laws to improve the conditions of the labour operation.

3.2.1 Types of Construction Labour

* The labour in construction industry can be broadly classified as :

  (a) Casual labour and (b) Labour on regular establishment.

* The casual labour is recruited according to the needs of construction industry. They are paid as per daily wages and payment is made at every week end or fortnight. Such labourers are not allowed any paid holiday or any other benefits.

* The labour on regular establishments are employed for longer periods. They are skilled workers and paid on monthly basis. They are entitled to holidays, leave and such other benefits.

3.2.2 Necessity of Labour Acts

1. To improve relations between the employees and the employer.
2. To help pay fair wages to workers.
3. To give compensation to workers, victims of accidents.
4. To reduce conflicts, strikes etc.
5. To procure job security for the workers.
6. To minimise unrest among the workers.
7. To promote wholesome environmental conditions in the industry or organisation.
8. To fix hours of working, rest, pauses etc.

3.2.3 Classification of Labour Laws

1. Laws regarding the working conditions of labour : Factory Act, 1948.
2. Laws regarding wages and other payments to labour : Minimum Wage Act, 1948.
3. Laws regarding the social securities of labour : Workers compensation Act, 1923.

3.2.4 Terms Used in the Main Provision of Act

1. **Adult :** A person who has completed 18 years of age.
2. **Adolescent :** A person who has completed 15 years of age but has not completed 18th year.
3. **Calendar year :** A period of 12 months, beginning from first of January.
4. **Child :** A person who has not completed 15 years of age.
5. **Day :** A period of hours, beginning at mid-night and ending at the next mid-night.
6. **Week :** A period of 7 days, beginning at mid-night on Saturday or any such night that may be approved in writing by the Chief Inspector of Factory.
7. **Power :** It is electrical energy or any other form of energy, which is mechanically transmitted.

8. Prime mover : It is an engine, motor or any other appliance that generates or otherwise provides power.

9. Machinery : It includes prime movers, transmission machinery and all other appliances, whereby power is generated, transformed, transmitted or applied.

10. Manufacturing process :

(a) It is any process for making, altering, repairing, finishing, packing, clearing, demolishing etc.

(b) Pumping oil water or sewerage.

(c) Constructing, reconstructing etc.

11. Worker : A person who is employed directly or through any agency.

12. Factory : Any premises :

(a) Wherein 10 or more workers are working or were working on any day of preceding 12 months and in any part of which, a manufacturing process is being carried on with the aid of power.

(b) Wherein 20 or more workers are working or were working on any day of the proceeding 12 months and in any part of which a manufacturing process is being carried out without the aid of any power.

13. Occupies : A person who has ultimate control over the affairs of factory.

Now, we will study some of the laws :

3.3 FACTORY ACT

- This law was passed in 1948. It is for the labour working in factories. Though it is not directly applicable to the labour working in construction industries yet the provisions are applicable to the labour convered under the term "Factories" as defined in the act such as these manufacturing fabricated components, form work fabrication, structural fabrication etc.

- **The law provides laying down the basic requirement such as :**

    1. Cleanliness, dispose of waste.

    2. Proper lighting and ventilation.

    3. Prevention of over-working.

    4. Provision of drinking water, hutments, latrines, urinals, washing places etc.

    5. Safety requirements.

    6. Welfare facilities.

    7. Fixation of working hours intervals for rest, paid holidays etc.

    8. Annual leave etc.

3.3.1 Main Features and Provisions of Factory Act

1. Approval, licensing and registration of the factory.

2. Health provisions :

(i) Cleanliness.

(ii) Ventilation.

(iii) Lighting.

(iv) Drinking water.

(v) Bath rooms.

(vi) Latrines and urinals.

(vii) Spitoons.

3. **Safety provisions :**

   (i)   Fencing of machinery.

   (ii) Casing of machinery.

   (iii) Hoists and lifts.

   (iv) Protection of eyes, feet.

   (v)   Precautions against dangerous fumes.

   (vi) Precautions against fire.

4. **Welfare provisions :**

   (i)   Washing facilities.

   (ii) Sitting facilities.

   (iii) First aid applications.

   (iv) Canteens.

   (v)   Shelter, rest room, lunch rooms.

   (vi) Welfare facilities.

5. **Hours of work :**

   (i)   Fixed hours of working as specified in writing by chief inspector.

   (ii) Holidays.

   (iii) Overtime wages.

   (iv) Restriction on double employment.

   (v)   Employment of female workers (day time duty only).

6. **Employment of young person :**

   (i)   No child who has not completed his 14th year shall be allowed to work in a factory.

   (ii) A child who has completed 14th year or an adolescent shall be allowed to work on producing a fitness certificate.

7. **Annual leave with wages :** The act provides a paid weekly holiday in a week. Besides this every worker who has completed 20 days of service in a calendar year, is entitled to get an earned leave with wages.

   (i)   One day per 20 days of actual work for adults.

   (ii) One day per 15 years of actual work for an adolescent.

8. **Dangerous operation :** If any operation carried out in a factory exposes any worker to a serious type of bodily injury, poisoning or disease then the State Government may declare that operation to be dangerous and may ask for proper safe guards or may prohibit such operation.

9. **Accidents and Diseases :** If in a factory, an accident results in death or causes any bodily injury, the manager shall send an intimation to the factory inspector within the prescribed time limit.

   When a worker suffers from any disease specified in the schedule, the manager shall send an information to the factory inspector.

10. **Penalties :** On any contravention of any of the provision of this Act or any Rule, manger of the factory shall be guilty of an offence and may be punished with imprisonment or fine as may be decided under this Act.

3.4 MINIMUM WAGES ACT

- Minimum Wagtes Act was passed in 1948 for the Welfare of labour and provides for fixing the minimum rate of wages of labour. The minimum wage provided should be sufficient to live accordingly to normal standards. It should also be enough to keep his body and soul together.

- Minimum wages differs from place to place according to local conditions. The Act aims at making provisions for the statutory fixation of the minimum rate of wages in a number of industries such as jinning press, textile mills, weaving rice, agriculture, building, construction etc. where there are extensive chances for the exploitation of labour.

- The minimum wages are fixed by the committee appointed by the Government. The payment of such wages are enforced by the chief inspector of factories. It is compulsory on the part of the employer to maintain a register as record of payment of minimum wages.

3.4.1 Important Aspects of Minimum Wage Act

The important aspect of the act are as follows

1. The minimum wage act lays down the fixation of –
   (i)   a minimum time rate of wages
   (ii) a minimum piece rate of wages
   (iii) a guaranteed time rate
   (iv) an overtime rate, for different occupation, classes of work, for children, adults, apprentices.
2. The minimum wages may consist of a basic rate of wages and a cost of living allowances to be fixed by the competent authority.
3. Fixing appropriate number of working hours in a day, weekly holiday and payment of overtime wages.
4. Maintaining register indicating particulars of employees, wages paid to them and receipt given by them etc.
5. The act authorises the appropriate Government to appoint inspectors for the purpose of listening and deciding the claims arising due to payment less than minimum wages.
6. Penalties may be imposed for violating the provisions of the act.

3.4.2 The Main Provisions of the Minimum Wage Act

(a) Fixation and revision of minimum wages :
1. Appropriate Government shall fix minimum rate of wages and shall review, at interval not exceeding 5 years. Thus, the minimum rate is fixed and revised.
2. An advisory board or committee may be appointed for fixing or revising the minimum rate of wages.
3. Different rates of wages are fixed for different scheduled employments, different classes of work and different for adult, children and apprentices. The minimum rate of wages may be fixed either by hour, by day, by month or by any longer period as may be prescribed.

(b) Payment of minimum rates of wages :
1. The minimum wages shall be paid as prescribed in cash.
2. An employer shall not pay less than the minimum rate of wages fixed by the Government.
3. The Government will have power to
   (i)   fix number of working hours in a day,
   (ii) provide a rest day in every seven days period of working,
   (iii) order payment for work on rest day not less than overtime rate.

(c) Maintenance of record :
- Employer shall maintain registers and other records giving details of employees, their nature of work, wages paid to them, their receipts etc. in the prescribed forms.

(d) Inspectors and their powers :
- Government may appoint inspectors to see that the provisions of the act are honoured. They shall have powers to enter the premises of the factories, check registers and records, examine any person and seize relevant documents in respect of an offence or breach of any provision.

(e) Claims :

- An employee himself or through official of recognised trade union or an inspector can apply to the proper authority i.e. labour commissioner to hear and decide the claims in respect of –

    (i) payment paid less than minimum wages rate

    (ii) wages of the overtime rate

    (iii) payment for day of rest or

    (iv) any other similar claims.

- The application for the claim should be presented with a period of six months from the date on which the claimed amount was payable. After receiving the application, the authority shall hear both the employee and the employer and decide the claim as per the set provisions in the act.

(f) Offences and penalties :

- The relevant provisions are as follows :

    1. An employer who contravenes any provision of the act is punishable and may be fined extending upto ₹ 500/-.

    2. An employer charged with an offence under this act can file a complaint against the person who has actually committed the offence and if so proved the actual offender shall be liable for the punishment.

    The law provides :

    1. Payment as per minimum wage and on working day.

    2. Payment on fixed date, normally on bazar days.

    3. The payment to be made without deductions except those items which are mentioned in the law.

3.5 WORKMAN'S COMPENSATION ACT 1923 (AS MODIFIED IN 1984)

- This most important law was first passed in 1923. Then in 1962, it was amended. The law protects the victims of accidents and their families from hardships arising out of, and in the course of employment.

- This Act is applicable to all the employees where wages are upto ₹ 1000/- per month and who are not engaged in clerical or administrative work. It is meant for the worker whose occupation is dangerous. Such a compensation is however not payable to disobedient workers or drunkards or consuming drugs or having gross negligence or wilful removal of safety guards.

- In the case of death, compensation is paid under all circumstances.

3.5.1 Main Factors of the Workman's Compensation Act

(a) The worker (or his dependent) can claim the compensation.

    In case of injury : If the injury has been caused by an accident in the course of employment.

(b) The amount of compensation depends upon the result of injury and the nature of the disablement.

(c) All fatal accidents are to be brought to the notice of the labour commissioner and the employer shall deposit the amount of compensation with the authority within thirty days.

However the employer is not required to pay the compensation if the injuries are due to

(i) negligence on the part of a worker by wilful disobedience of safety and security regulations.

(ii) Non-observance of safety measures and not using the safety guards.

(iii) influence of liquor or drugs.

(iv) diseases which are not caused as a result of working on the job.

3.5.2 Causes of Accidents

Accidents may result due to :

(i) Mechanical break down.

(ii) Absence or failure of safety devices.

(iii) Live electrical wires.

(iv) Faulty lifting and transporting equipments

(v) Unsafe scaffolding and shuttering etc.

(vi) Human causes such as poor eye sight, fatigue, carelessness and effect of intoxicants.

(vii) Environmental factors such as inadequate space, poor lighting and ventilation etc.

3.5.3 Important Definitions in Respect of Compensation Act

(a) Dependent : They include widow, a minor son, unmarried daughter or a widowed mother and if wholly or partly dependent on the earnings of the workman at the time of death : a minor brother or unmarried sister, a widowed daughter in law, a minor child of predeceased daughter etc.

(b) Partial disablement : It is disablement of temporary nature and which reduces the earning capacity of a workman.

(c) Total disablement : It is the disablement which temporarily or permanently incapacitates a workman for all work and he cannot earn at all for some period or forever.

(d) Workman : He is a person who is (i) a regular servant employed in any such capacity specified in schedule and not permanently employed in any administrative capacity (ii) getting wages not exceeding ₹ 1000/- p.m.

3.5.4 Employers' Liability for Compensation

The employer is liable for compensation in the following cases :

(i) If the injury has been carried by an accident during the course of employment.

(ii) The injury has resulted in workman's death, permanent or temporary or total or partial disability.

(iii) If any disease contacted in an occupational disease peculiar to the employment such as silicosis to foundry worker, caison sickness for a worker working in pneumatic caisson, the employer shall give compensation to the worker.

3.5.5 Amount of Compensation

The different factors which are considered in deciding the amount of compensation are as follows :

1. Average monthly wages of the workman which should not be more than ₹ 1000/-.

2. The extent of injury.

3. The nature of injury e.g. (i) Death, (ii) Permanent total disability, (iii) Permanent partial disability, (iv) Total or partial temporary disability.

The amount of compensation payable in case of different injuries are as per schedule IV of the act.

3.5.6 Distribution of Compensation

The payment of compensation in respect of death or any legal disability shall be deposited with the commissioner. The employer can give advance to any dependent on account of compensation not exceeding ₹ 100/-. The commissioner, after deducting the advance of any, may allot the entire amount of the compensation to any one of the dependents.

3.5.7 Notice for Claim of Accident

A workman injured in an accident has to give a notice in writing of the accident to the employer. It should contain all the particulars of the workman injured and of its cause so that the employer can check the facts. A claim for the compensation must be made within period two years of the occurrence of the accident/ death or in case of occupational disease within two years from the day workman feels the notice of displacement.

3.5.8 Medical Examination

(i) When a workman has given notice of an accident, he shall offer himself to be examined free of charge by a qualified medical practitioner before the expiry of 3 days from the time at which the notice has been issued.

(ii) If a workman refuses to submit himself for examination by a qualified medical practitioner, his right to compensation shall be suspended.

(iii) If workman dies without having been medically examined, the payment of compensation can be claimed-by the dependents of deceased workman.

3.5.9 Appointment of Commissioners

The act provides for the appointment of commissioners for resolving disputes arising out of claims for compensation. All payments of compensation are made through the commissioner. Every commissioner shall be deemed to be public servant within the meaning of the Indian Penal Code.

Important Points

- Safety Engineering "is a term for all efforts on the part of the management to prevent accidents. Safety in construction is considered important.

- **Definition of Safety :**

  It involves a three pronged requirements.

  1. Freedom from danger or risk.

  2. Freedom from injury.

  3. Harmlessness which is the state of being without hurt, loss or liability.

- I.F. Rate $= \dfrac{\text{Numer of disabling injuries} \times 1000000}{\text{Number of man hours worked}}$

- I.S. Rate $= \dfrac{\text{Number of days lost} \times 1000}{\text{Number of man hours worked}}$

- **Management's Role :** Top management must also be interested in the safety programmes. Their involvement is equally necessary for implementing safety programmes. In fact, the programme has to originate at the top without which it will not be a success. The safety programme must be given publicity.

- A ladder should be placed at an angle of 75° to the horizontal. The foot is to be tied so as to be driven into the ground or placed on a sand bag or sole plate to prevent the base from going sideways. A ladder should never be placed on a pile of loose bricks, oil drum or dust bin.

- Injuries are caused by falls from eaves or gutter levels. Falls from eaves level can be prevented by providing a barrier on the roof slope at the eaves. The barrier must extend to a sufficient height and be strong enough to stop a man sliding down the roof slope.

- The condition of the labour working in construction industry is very poor. The number of workers employed in construction industry is nearly four crores. They are mostly unskilled or semiskilled workers. In construction industry, the contractor dictates the terms of service conditions particularly to skilled workers. They are paid very poor wages.

- **Necessity of Labour Acts**

  1. To improve relations between the employees and the employer.

  2. To help pay fair wages to workers.

  3. To give compensation to workers, victims of accidents.

  4. To reduce conflicts, strikes etc.

  5. To procure job security for the workers.

  6. To minimise unrest among the workers.

  7. To promote wholesome environmental conditions in the industry or organization.

  8. To fix hours of working, rest, pauses etc.

- **Laying down the basic requirement such as :**
  1. Cleanliness, dispose of waste.
  2. Proper lighting and ventilation.
  3. Prevention of over-working.
  4. Provision of drinking water, hutments, latrines, urinals, washing places etc.
  5. Safety requirements.
  6. Welfare facilities.
  7. Fixation of working hours, intervals for rest, paid holidays etc.
  8. Annual leave etc.

- **Minimum Wage Act :** The important aspect of the act are as follows :

  The minimum wage act lays down the fixation of –
  - (i) a minimum time rate of wages
  - (ii) a minimum piece rate of wages
  - (iii) a guaranteed time rate
  - (iv) an overtime rate, for different occupation, classes of work, for children, adults, apprentices.

- **Workman's Compensation Act 1923 (As modified in 1984) :** This most important law was first passed in 1923. Then in 1962, it was amended. The law protects the victims of accidents and their families from hardships arising out of, and in the course of employment.

- When a workman has given notice of an accident, he shall offer himself to be examined free of charge by a qualified medical practitioner before the expiry of 3 days from the time at which the notice has been issued.

Practice Questions

1. State the meaning of "Safety Engineering". Give the importance of safety in Civil Engineering construction works.
2. State the causes of failure leading to damage to life and property.
3. Mention the important points, the construction engineer should bear in mind in order to ensure safety.
4. Name four jobs in construction requiring attention for ensuring safety.
5. State the tangible and intangible costs with reference to an accident on construction site.
6. Give the safety programme which should be introduced to avoid accidents.
7. State how do you ensure safety in the following works : (i) Excavation, (ii) Scaffolding, (iii) Roofing.
8. Explain the terms : (i) Injury Frequency Rate, (ii) Injury Severity Rate.
9. A contractor employed 200 men for sixty weeks for construction work, the working hours being 40 hours per week. Six disabling injuries occurred during this period. What is the Injury Frequency Rate ?
10. State the general rules that must be abided by the contractor with respect to :
    - (i) Minimum Wages Act, and
    - (ii) Workmen's Compensation Act.
11. Write down the steps you think necessary to improve the working conditions of construction labour in India.
12. Distinguish between Minimum Wages Act and Payment of Wages Act.
13. State the provisions made in the "Factory Act".
14. State the provisions made in "Minimum Wages Act".
15. Explain the main provisions of the workmen's compensation Act ?

EXPERIMENTS

Experiment No. 1

Aim : Prepare the organization chart of any one government/public sector organization executing any major civil engineering projects in your area.

Public Works Department (P.W.D.) :

P.W.D. has history of over 150 years in state of Maharashtra. This department deals with the construction and maintenance of roads, bridges and government buildings. It also acts as the technical advisor to the State Government of Maharashtra.

General Organizational Set-up :

For ensuring effective timely completion of P.W.D.'s work, department has evolved a specific set up of organization with following wings :

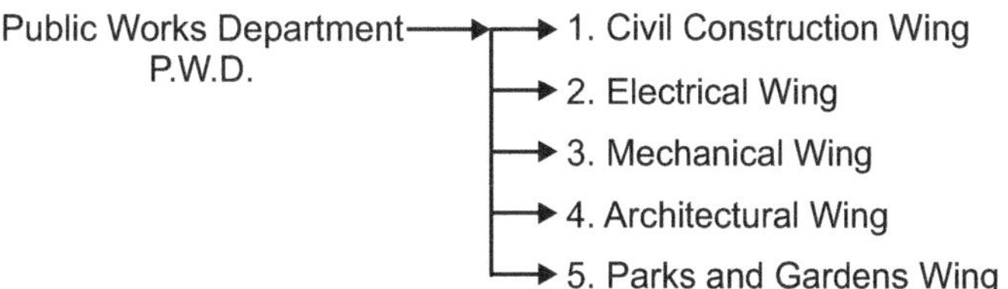

The control of operations of all wings vests in Government of Maharashtra by the Minister (Public works) and Minister of state (Public works) and they are assisted by secretary (works) and secretary (Roads).

Civil Construction Wing

Chief Engineer

(For Mumbai, Nagpur, Pune, Aurangabad, Nashik and Amravati region, one for each region in addition to this two chief engineers are posted to Mumbai)

↓

Superintending Engineer

↓

Executive Engineer

↓

Deputy Engineer

↓

Sectional Engineer

❑❑❑

Experiment No. 2

Aim : Prepare the organization chart of any one private organization executing any major civil engineering projects in your area.

Organizations play important role in the development of the nation. The construction organizations are important in the role of development, by providing one of the three basic needs of humankind - A house. Also provide civilized area of country.

Any organization develops from number of stages, from foundation till successful completion. In private civil engineering organizations, the organization is a centralized type of organization. Here, in this type, individual departments with their heads report to the top management.

Normally, a construction organization consists of the following sections or departments :

1. Engineering department.
2. Purchase department.
3. Finance department.
4. Administrative department.
5. Sales and marketing department.
6. Legal department.
7. Accounts department.
8. Electronic Data Processing (E.D.P.) department.

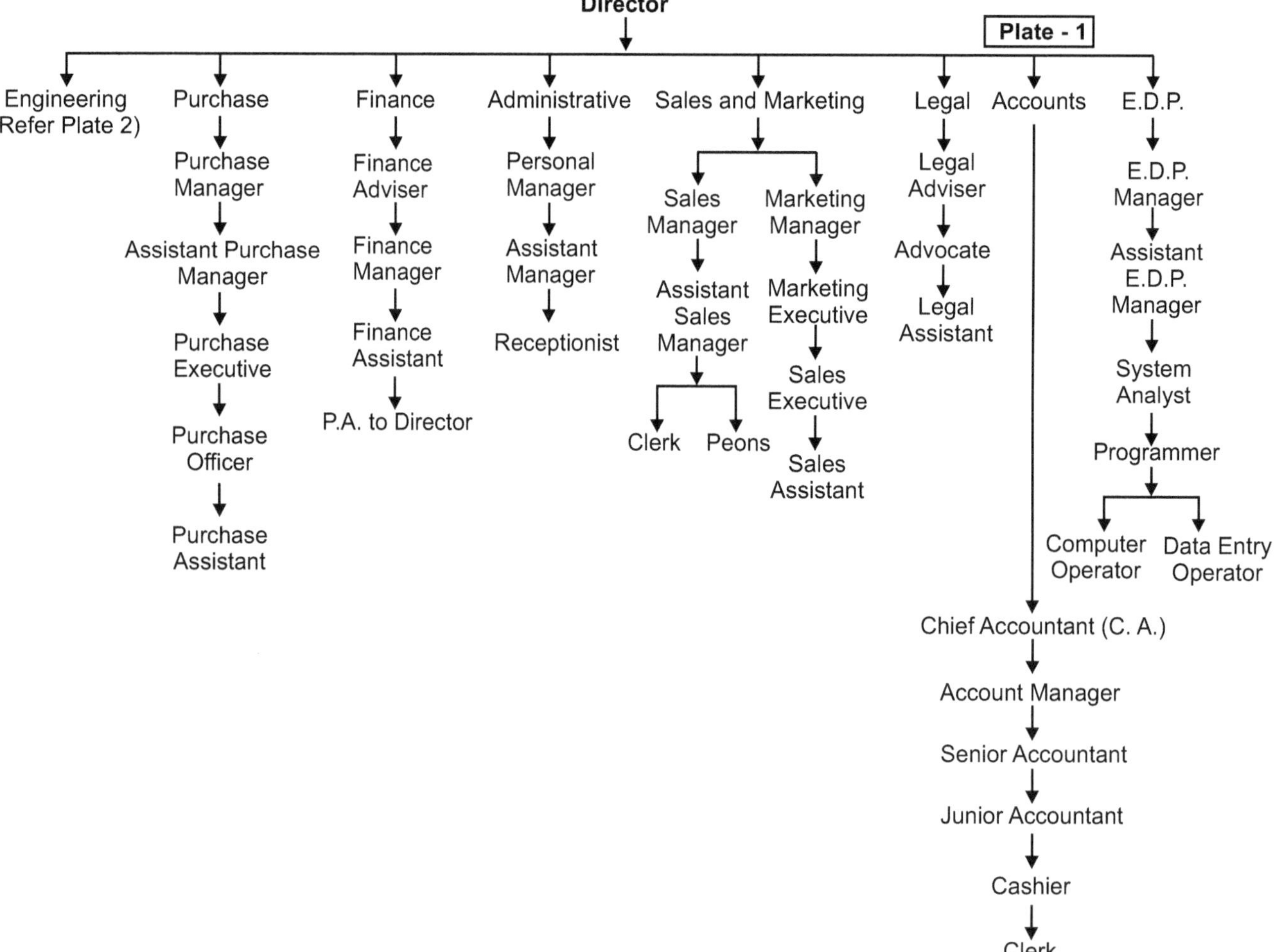

Engineering Department

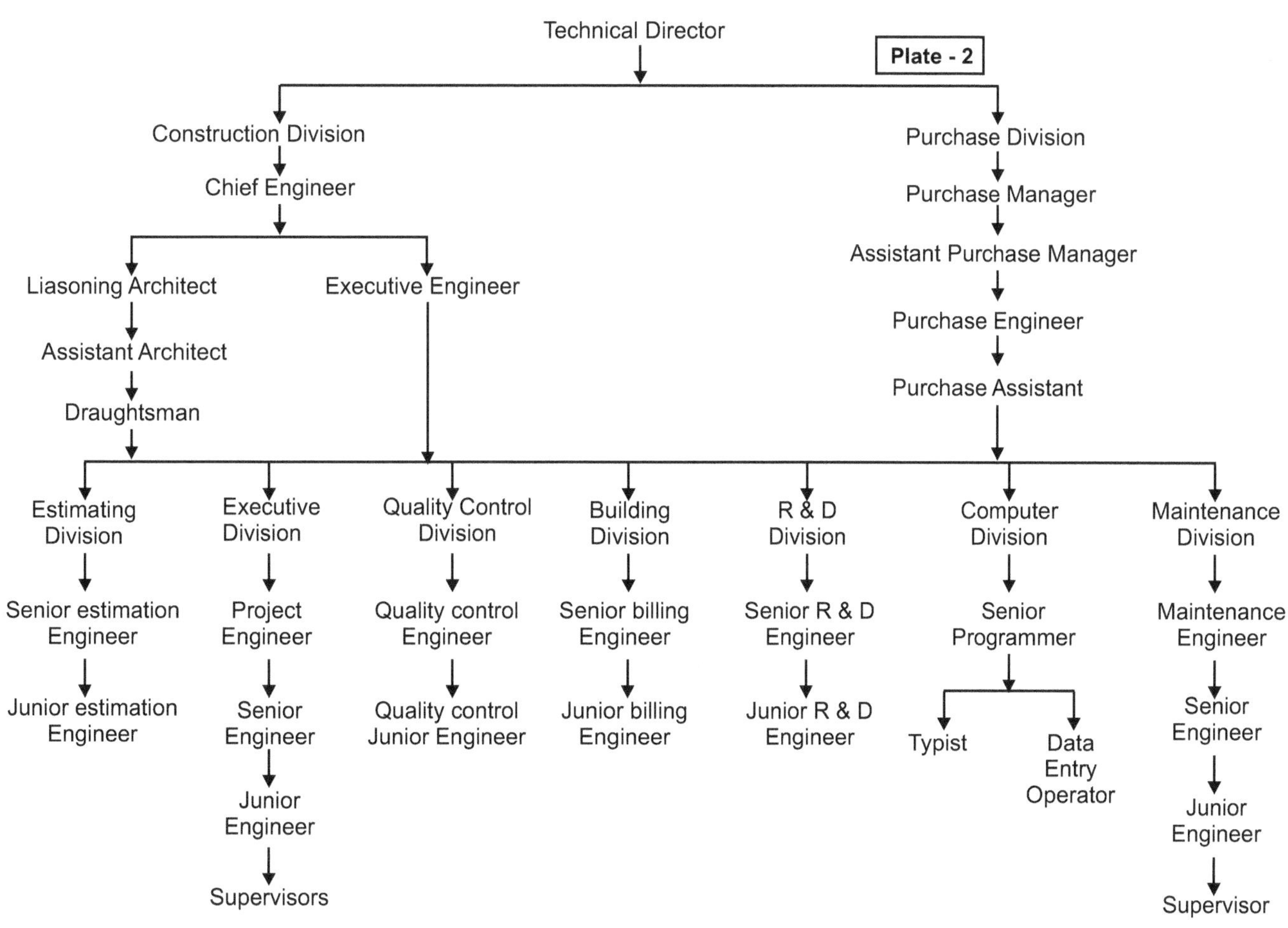

❑❑❑

Experiment No. 3

Practical Outcome (Pro) :

Aim : Prepare the list of roles and responsibilities of various personnel in any government construction organization.

Officers in Civil Wing of P.W.D. :

Civil construction wing of the Public Works Department is in following four cadres as per hierarchy.

1. Chief Engineer, P. W. Region.
2. Superintending Engineer, P. W. Circle.
3. Executive Engineer, P. W. Division.
4. Deputy Engineer, P. W. Sub-division.

1. Chief Engineer :

The chief engineer is the professional head of the department and is responsible to Government for efficient working.

The Chief Engineer shall :

(i) Provide technical guidance and guidance for supervision for ensuring both quality and progress of work.

(ii) Ensure co-ordination of activities of different wings of the department.

(iii) Carry out inspections of the circle and divisional officers under his control as per following frequency.

    (a) Circle offices under his control once every year.

    (b) Divisional offices - Ten divisional offices each year or every divisional office in every three to four years.

(iv) Control over the performance of the officers maintaining the accounts.

(v) Monitor the budget allocations in general and advise the superintending engineers to avoid excess or lapse of grants.

(vi) If in exercising the control, any financial irregularity is discovered and if it considers to be serious as to require disciplinary action.

2. Superintending Engineer :

The superintending engineer is the administrative head of the circle and is responsible to the chief engineer and looks after the administration and general professional control of the in-charge officers of the department within the circle.

The Superintending Engineer shall :

(i) Inspect the various works within circle and satisfy that the system of management prevailing is efficient and economical.

(ii) Ascertain the efficiency of the subordinate offices and outdoor establishment and report whether the staff employed in each division is actually necessary and adequate for its management.

(iii) Carry out inspections of the divisional offices under control once every year and those of sub-divisional offices under, once in five years.

(iv) The superintending engineer is the controlling authority for budget and sees that the accounts for works, stocks, tools and plants are maintained throughout circle and the executive engineers submit the accounts to the accountant general punctually.

(v) Exercise the control on monthly expenditure as per the monthly cash-flow system introduced.

3.　Executive Engineer :

The executive unit of the Department is the division in charge of an executive engineer who is responsible to the Superintending Engineer for the execution and management of all works within the division.

The Executive Engineer shall :

(i)　See that proper measures are taken to preserve all the buildings and roads in his division and to prevent encroachment on Government lands in charge.

(ii)　Keep accurate plans of all Government lands.

(iii)　Act as ex-officio professional adviser to all the departments of the administration within the limits of charge.

(iv)　Carry out inspection of the sub-divisional officers under control once every year.

(v)　Commences and construction of any work or spends funds after receipt of the sanction orders of the competent authority.

(vi)　For the technical matters in division office, Deputy Executive Engineer, three to four Junior Engineers, Draftsman, Tracer etc. assist the Executive Engineer.

(vii)　For the administrative matter in division office, first clerk assist the Executive Engineer.

(viii)　For the accounts matter, divisional accountant and for purchase of material etc., storekeeper assist the Executive Engineer.

4.　Deputy Engineer (Dy. Engineer) :

A division is divided into sub-divisions in change of an Assistant Executive Engineer class-I/Assistant Engineer Grade-I/Sub-divisional engineer/Sub-divisional officer/Deputy engineer.

The Deputy Engineer shall :

(i)　Responsible to the Executive Engineer in charge of the division for the management and execution of works within sub-divisions.

(ii)　Ensure proper and timely execution of work.

(iii)　Ensure quality and quantity of work through testing of materials and work done at specified intervals/frequencies.

(iv)　Prepare estimate/revised estimate/completion reports timely and exercises control on account pertaining to the division and submits the same timely to the Executive Engineer.

(v)　Executes works related to E.G.S., M.L.A., M.P. Fund works, deposit works, etc.

(vi)　For the technical matters in sub-divisional office, 5 to 6 Junior Engineers assist Dy. Engineer. Preparation of works estimate, supervision of work, recording measurement of the works are the important duties of Junior Engineer.

(vii)　For the administrative matter, account and store matters, senior clerk assist Dy. Engineer of the sub-division.

❑❑❑

Experiment No. 4

Practical Outcome (Pro) :

Aim : Prepare the list of roles and responsibilities of various personnel in any private construction organization.

1. **Chief Engineer shall :**
   (i)   Survey the various projects before execution starts.
   (ii) Finalize the specifications as per the market and sales requirement.
   (iii) Prepar financial schedule.
   (iv) Finalize other developmental works.
   (v)   Be responsible to higher authority director.
   (vi) Give all technical details to Executive Engineer to execute.

2. **Liasoning Architect shall :**
   (i)   Prepare the plans and drawings as per the requirements of the market and sales requirement.
   (ii) Assistant architects and draftsman assist the Liasoning architect in technical matter.
   (iii) Co-ordinate with structural engineer for final structural drawings alongwith specifications and requirements.

3. **Estimation Engineer shall :**
   (i)   Prepare estimate, rate analysis as per drawings and specifications received.
   (ii) Revise the estimate if any revision is made in the given details.
   (iii) Maintain the records for future reference.

4. **Project Engineer shall :**
   (i)   Prepare job layout, breakdown structure of given project.
   (ii)   Prepare schedule of work (bar chart) for given project.
   (iii) Prepare material schedule and submit to purchase department.
   (iv)   Update the schedule as and when required.
   (v)   Entrust works assigned to contractors and labours.
   (vi)   Procure various materials, tools and plants as and when required.
   (vii) Supervise and execute the work of given project.

5. **Quality Control Engineer shall :**
   (i)   Check the quality of work going by inspection.
   (ii) Check the quality of materials before using.
   (iii) Entrust the quality of materials purchased by vendors.
   (iv) Prepare the records of assessment of materials purchased/procured and maintain them.

6. **Billing Engineer shall :**
   (i)   Check the bills as per estimate prepared and sanctioned.
   (ii) Prepare the bills of works done by contractors.
   (iii) Passing the bills for further assistance and forwarding to accounts department for payment.
   (iv) Maintain the records of all bills for future reference.

7. **Maintenance Engineer shall :**
   (i)   Maintain the work executed.
   (ii) Supervise the work maintained by labours under guidance.
   (iii) Prepare schedule of routine maintenance to assure the quality of work.
   (iv) If any flaws found during routine maintenance, report to quality engineer and executive engineer.

Experiment No. 5

Practical Outcome (Pro) :

Aim : Prepare the bar chart for given construction project.

- **Bar chart :**

Bar chart is a graphical representation of construction work and time required.

It is also called as Gantt bar chart.

It is most commonly used method in major construction industries.

- **Data required for preparation of bar chart :**

Following basic data is required to be collected before preparation of bar chart :

(i)　Total time available for any particular project or activity.

(ii)　Availability of funds.

(iii)　Availability of labours or requirement of labours for each activity.

- **Usefulness of bar chart :**

Following are the uses of bar chart :

(i)　Total time required for any particular activity.

(ii)　Schedule of required materials.

(iii)　Labours required for any particular activity.

(iv)　Preparation of work before each activity.

(v)　Progress of work updation.

(vi)　Total finance requirement schedule.

(vii)　Preparation of other activities simultaneously.

(viii)　Advance planning of any particular activity, for availability of materials and purchasing new materials.

(ix)　Following up with other agencies or contractors.

(x)　Reasons for delay for any activity.

Example :

1.　Prepare bar chart of a particular building/project from commencement right till completion.

Example :

2.　For a R.C.C. building, maintenance work is going on with new masonry work till finishing. Prepare a bar chart for such case for all further activities according to the scheduled time.

Example 1 :   **DATE :** ......

NAME OF THE PROJECT

Bar Chart and Monthly Cash Flow for New Building

Sr. No.	Activity	JAN	FEB	MAR	APR	JUN	JUL	AUG	SEP	OCT	NOV	DEC	JAN	FEB	MAR	APR	Total
1.	Levelling and cleaning. Initial development	▬															2
2.	Plinth work		▬														8
3.	First slab			▬													4
4.	Second slab				▬												5
5.	Third slab					▬											5
6.	Fourth slab						▬										5
7.	Fifth slab							▬									5
8.	Terrace slab								▬								5.5
9.	Overhead water tank, lintels and lofts										▬						3
10.	Masonry work					▬▬▬▬▬▬											11
11.	Doors and Windows					▬▬▬▬▬											8
12.	Plastering						▬▬▬▬▬▬										12
13.	Flooring								▬▬▬▬▬▬								5
14.	Plumbing						▬▬▬▬▬▬▬▬										4
15.	Electrical work						▬▬▬▬▬▬▬▬										3
16.	Waterproofing							▬▬▬▬▬									6
17.	External development											▬▬▬▬▬					5
18.	Painting													▬▬▬			2.5
19.	Possession															▬	1
	Cash flow/month in %	2	8	5	7	12.5	14	10	10	6	7	6	4	3	3	2.5	100

Example 2 :   **DATE :** ....

NAME OF THE PROJECT

Bar Chart and Monthly Cash Flow for Maintenance Renovation of Building

Sr. No.	Activity	PERIOD IN MONTHS					
		JAN	FEB	MAR	APR	MAY	JUNE
1.	Masonry work	▬					
2.	Window fixing	▬					
3.	External plaster	▬▬▬	▬				
4.	Internal plaster		▬▬	▬			
5.	Flooring			▬			
6.	Floor finishing				▬		
7.	W.C./Bath waterproofing	▬	▬▬	▬▬	▬		
8.	Plumbing				▬		
9.	Electrification				▬	▬	
10.	Dado				▬	▬	
11.	Kitchen platform work				▬	▬	
12.	Tile polishing				▬	▬	
13.	Terrace waterproofing				▬		
14.	Staircase tilting					▬	
15.	Painting					▬	▬
16.	Carpentry				▬	▬	
17.	Electrical switch fixing					▬	
18.	Sanitary fittings fixing					▬	▬
19.	Glass fixing					▬	
20.	Drainage line connection				▬	▬	
21.	Cleaning and possession						▬

❏❏❏

Experiment No. 6

Practical Outcome (Pro) :

Prepare a network for given construction project to identify the critical activity in a project (to develop the critical path).

Example 1 :

From the data table, prepare the network diagram, define the completion period and select critical path.

Activity	Duration in days	Activities immediately	
		Preceding	**Following**
A	3	None	B, C
B	4	A	D
C	6	A	D, E
D	3	B, C	F
E	6	C	G
F	4	D	I
G	5	E	H, J
H	3	G	I
I	6	F, H	L
J	4	G	K
K	4	J	L
L	4	I, K	None

Solution :

Activity	Arrow	Duration	EST	LST	EFT	LFT	TF	FF	IF	Remarks
A	1 - 2	3	0	0	3	3	0	0	0	Critical
B	2 - 4	4	3	12	7	16	9	2	7	–
C	2 - 3	6	3	3	9	9	0	0	0	Critical
D	4 - 5	3	9	16	12	19	7	0	7	–
E	3 - 6	6	9	19	15	15	0	0	0	Critical
F	5 - 8	4	12	19	16	23	7	7	0	–
G	6 - 7	5	15	15	20	20	0	0	0	Critical
H	7 - 8	3	20	20	23	23	0	0	0	Critical
I	8 - 10	6	23	23	29	29	0	0	0	Critical
J	7 - 9	4	20	21	24	25	1	0	1	–
K	9 - 10	4	24	25	28	29	1	1	0	–
L	10 - 11	4	29	29	33	33	0	0	0	Critical

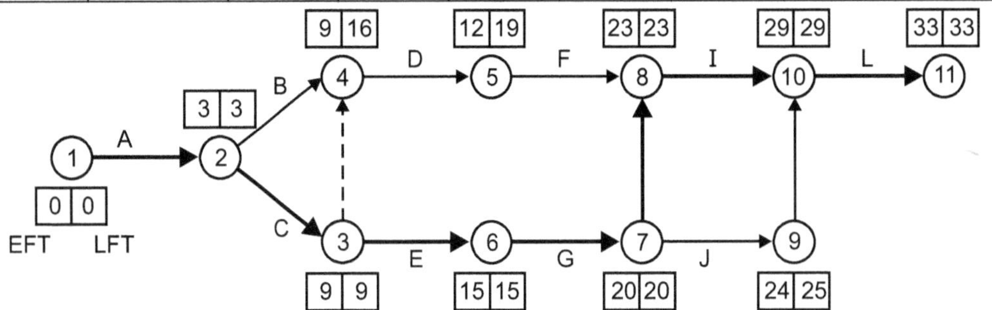

From figure it is seen that the total completion time period is 33 days.

Activities joining the events 1, 2, 3, 6, 7, 8, 10 and 11 is critical path and A-C-E-G-H-I-L are the critical activities.

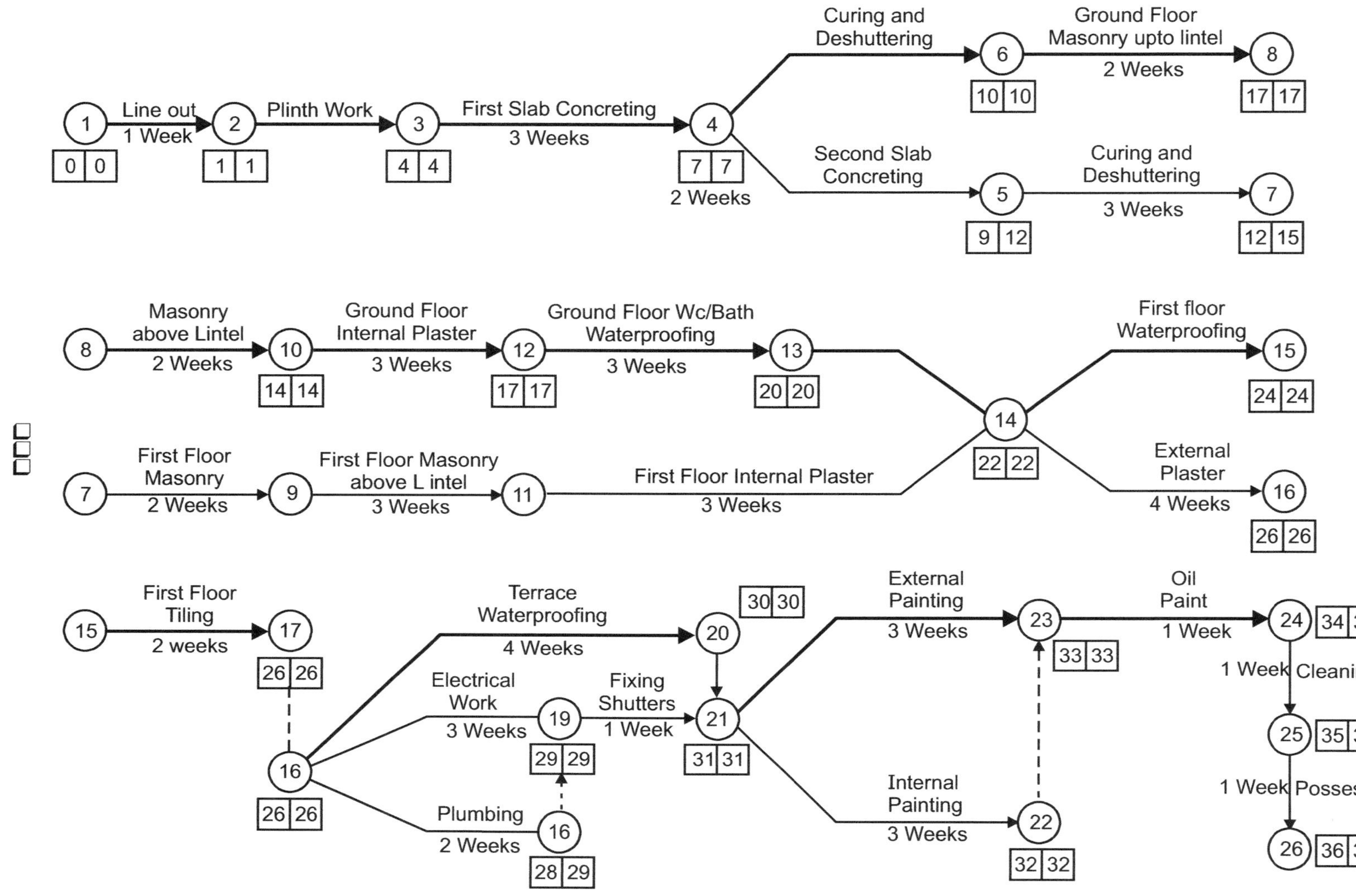
Line out
1 Week
Plinth Work
First Slab Concreting
3 Weeks
Curing and
Deshuttering
Ground Floor
Masonry upto lintel
2 Weeks
Second Slab
Concreting
2 Weeks
Curing and
Deshuttering
3 Weeks
Masonry
above Lintel
2 Weeks
Ground Floor
Internal Plaster
3 Weeks
Ground Floor Wc/Bath
Waterproofing
3 Weeks
First floor
Waterproofing
First Floor
Masonry
2 Weeks
First Floor Masonry
above L lintel
3 Weeks
First Floor Internal Plaster
3 Weeks
External
Plaster
4 Weeks
First Floor
Tiling
2 weeks
Terrace
Waterproofing
4 Weeks
Electrical
Work
3 Weeks
Fixing
Shutters
1 Week
Plumbing
2 Weeks
External
Painting
3 Weeks
Internal
Painting
3 Weeks
Oil
Paint
1 Week
1 Week Cleaning
1 Week Possession

Experiment No. 7

Practical Outcome (Pro) :

Determination of EOQ (Economic Order Quality) based on the given data.

Example 1 :

Determine the Economic Order Quantity from the following data :

Annual consumption of cement bags = 15000 units (per annum).

$$\text{Cost per bag} = ₹\ 370/\text{-}$$

Cost of placing order and processing the delivery = ₹ 15/- per order.

$$\text{Inventory carrying cost} = 15\% \text{ of unit value.}$$

Given data :

$$\text{Annual consumption} = A = 15000 \text{ units}$$
$$\text{Ordering cost} = B = ₹\ 15/\text{- per order}$$
$$\text{Cost per unit} = C = ₹\ 370/\text{-}$$
$$\text{Inventory carrying cost} = S = 15\% \text{ of unit value}$$

Solution :

$$\text{Economic Order Quantity} = EOQ = \sqrt{\frac{2AB}{CS}}$$

$$= \sqrt{\frac{2 \times 15000 \times 15 \times 100}{370 \times 15}}$$

$$= 90.05 \text{ units}$$

Example 2 :

Calculate the Economic Order Quantity for row materials and packing materials with following data :

Cost of ordering : Raw materials = ₹ 1000/- per order

$$\text{Packing materials} = ₹\ 5000/\text{- per order}$$

Cost of holding inventory : Raw materials : 1 paisa per unit per month.

Packing materials : 5 paisa per unit per month.

Production rate : 200000 units per month.

Solution : (A) For raw materials

$$\text{Economic Order Quantity} = EOQ = \sqrt{\frac{2AB}{CS}}$$

$$= \sqrt{\frac{2 \times 200000 \times 1000 \times 100}{1}}$$

$$= 200000 \text{ units}$$

(B) For raw materials :

$$\text{Economic Order Quantity} = EOQ = \sqrt{\frac{2AB}{CS}}$$

$$= \sqrt{\frac{2 \times 200000 \times 5000 \times 100}{5}}$$

$$= 200000 \text{ units}$$

❑❑❑

Experiment No. 8

Aim : Compute activity time, event time and floats for network drawn in practical no. 6.

Example 1 :

From the data table, prepare the network diagram, define the completion period and select critical path.

Activity	Duration in days	Activities immediately	
		Preceding	**Following**
A	3	None	B, C
B	4	A	D
C	6	A	D, E
D	3	B, C	F
E	6	C	G
F	4	D	I
G	5	E	H, J
H	3	G	I
I	6	F, H	L
J	4	G	K
K	4	J	L
L	4	I, K	None

Solution :

Activity	Arrow	Duration	EST	LST	EFT	LFT	TF	FF	IF	Remarks
A	1 - 2	3	0	0	3	3	0	0	0	Critical
B	2 - 4	4	3	12	7	16	9	2	7	–
C	2 - 3	6	3	3	9	9	0	0	0	Critical
D	4 - 5	3	9	16	12	19	7	0	7	–
E	3 - 6	6	9	19	15	15	0	0	0	Critical
F	5 - 8	4	12	19	16	23	7	7	0	–
G	6 - 7	5	15	15	20	20	0	0	0	Critical
H	7 - 8	3	20	20	23	23	0	0	0	Critical
I	8 - 10	6	23	23	29	29	0	0	0	Critical
J	7 - 9	4	20	21	24	25	1	0	1	–
K	9 - 10	4	24	25	28	29	1	1	0	–
L	10 - 11	4	29	29	33	33	0	0	0	Critical

Experiment No. 9

Aim : Check the breakdown structure of the given typical construction project to justify its role in managing its relevant activities.

Work Breakdown Structure (WBS) Development Procedure :

Project WBS development is based on a technique called decomposition. Decomposition provides division of work into smaller, more manageable components. It is arranged in a hierarchical order.

WBS Development Involves the following Major Steps :

Step 1 : Identification of the program in the projects, if any.

The project is to be defined in terms of independent major physical products of the program.

Step 2 : Identification of the major deliverables of each project.

The major deliverables should always be defined in terms of actual management and execution.

There are two types of approach :

1. **Noun-type approaches :**

    Noun-type approaches define the deliverable of the project work in terms of the physical components that make up the deliverable.

    For example : The physical units (areas - A1, A2, A3; sections - S1, S2, S3; building blocks - Admin, CED, MED, Library)

2. **Verb-type approaches :**

    Verb-type approaches define the deliverable of the project work in terms of the actions that must be done to produce the deliverable.

    For example : The phases of the project life cycle (design, material procurement, construction, possesion), or the work speciality (civil, mechanical, HVAC, etc.).

Step 3 : Decomposition of major deliverables.

It is to be done in appropriate detail manner to achieve integrated control and proper management.

Check whether the proper estimate of costing and duration of activities can be developed at this step for each deliverable.

For each deliverable, if there is adequate detail, proceed to step 5, if not proceed to step 4, which means that different deliverables may have different levels of decomposition.

Step 4 : Identification of constituent components of each deliverable.

As with deliverables, sub-deliverables should be described in terms of tangible, verifiable results to facilitate performance measurement. Divide until criteria in step 3 is achieved. Now you are at work package level. Every effort should be made to develop a WBS that is output-oriented in order to concentrate on concrete deliverables. If a WBS follows the organizational structure, the focus will be on the organization function and processes rather than the project output or deliverables.

A Work Breakdown Structure is normally represented as per following chart :

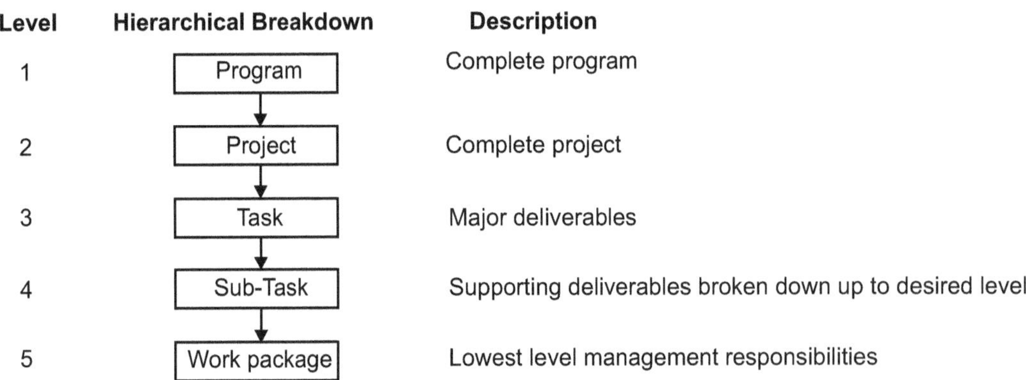

Hierarchial chart for WBS

Example :

Prepare a Work Breakdown Structure for a Stilt+7 residential building.

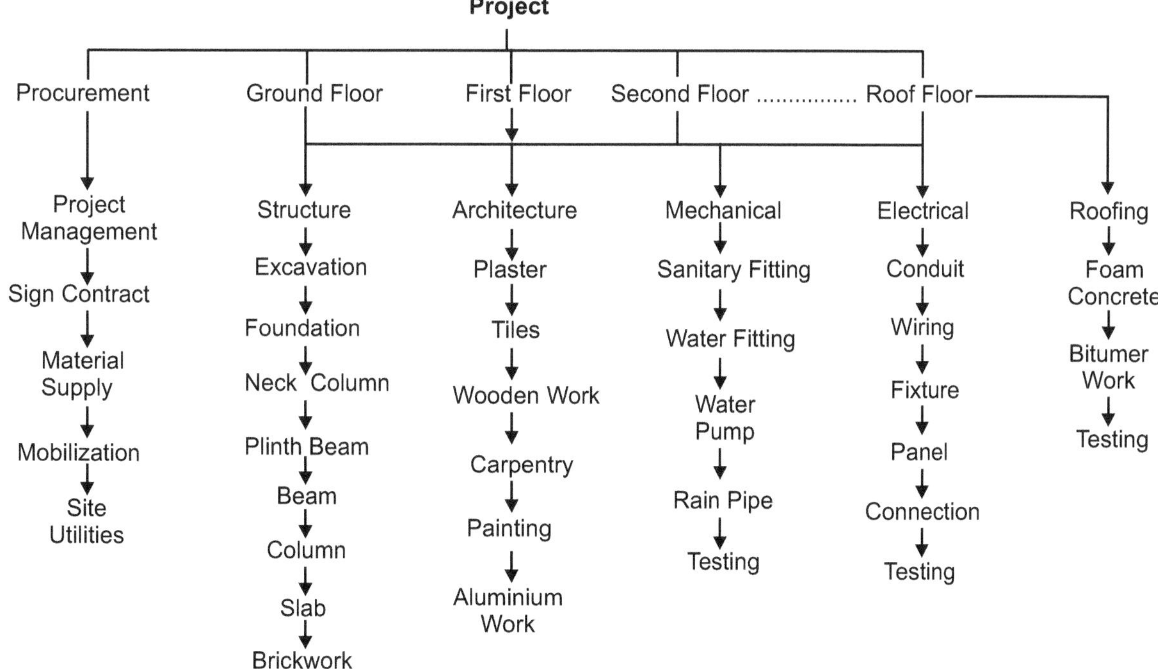

Work Breakdown Structure

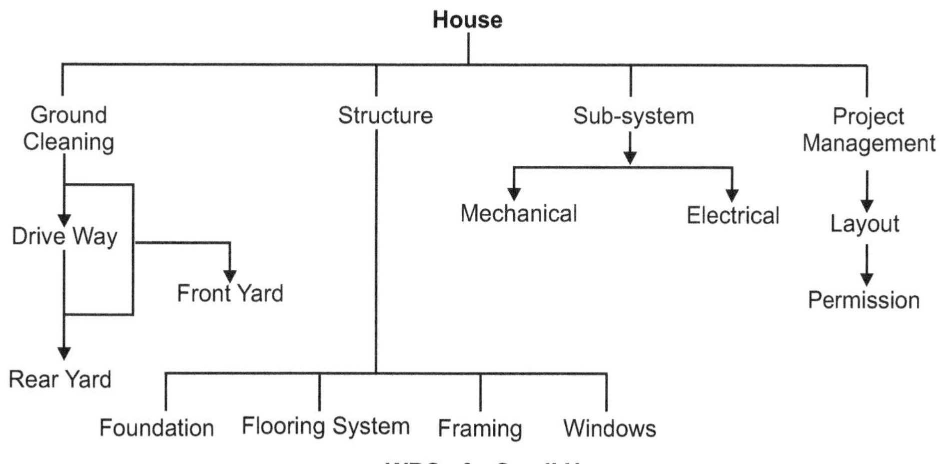

WBS of a Small House

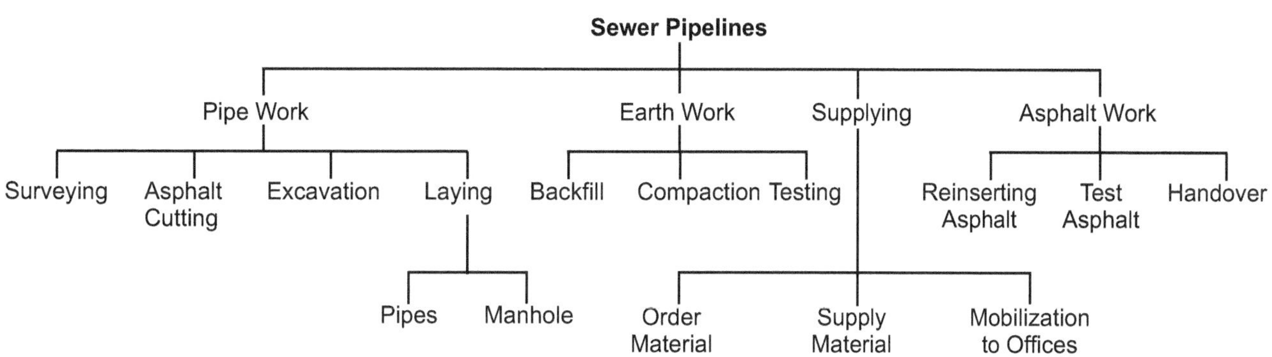

WBS of Roadwork with Pipelining

❏❏❏

Experiment No. 10

Aim : Prepare Job layout for the given construction site.

A Job layout is planning and optimization of site which can reduce the transportation flows, moving, reloading and enhancing the services at the construction work, to improve the work productivity and thus the costs of a project.

Due to the complication of the site layout planning problem, construction managers often perform this task of site layout planning by using thumb rule, ad-hoc rules, and first-come-first-serve approach which leads to confusion and even to inefficiency. But improper management and planning leads to risk to labours safety, construction material safety and may increase the overall cost of project. To avoid such problems job layout should be performed before execution of any construction project.

Example :

Prepare a Job Layout for a residential building having 7 buildings of stilt+4 level.

Site Plan :

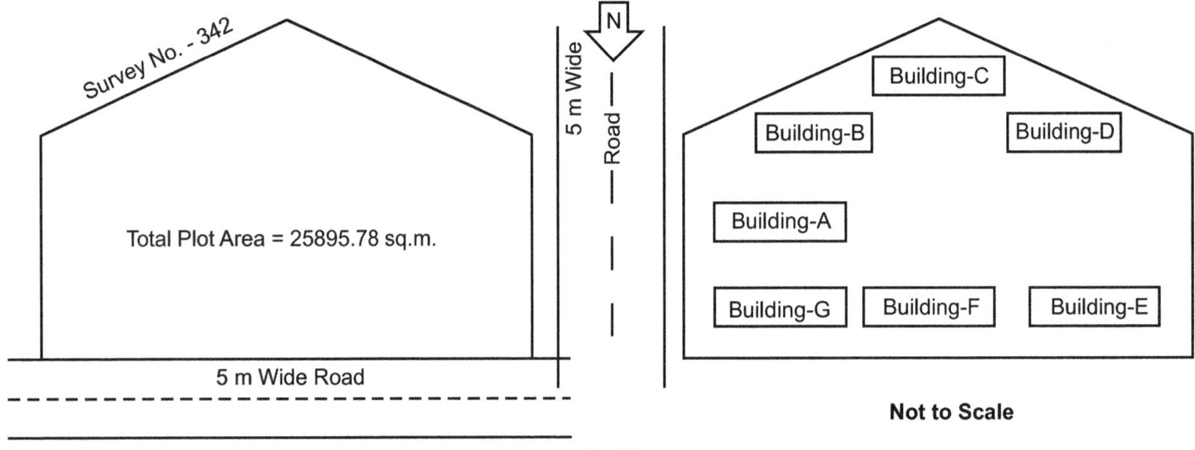

Site Plan

Step 1 : Identification of different facilities to be provided.

The following Temporary Facilities are identified to be provided at site :

1. Site office
2. Booking office
3. Subcontractor's office
4. First aid and medical room
5. Guard room
6. Toilet on site
7. Engineer and Staff quarters
8. Labour quarters
9. Equipment maintenance room
10. Parking for machines
11. Bar bending shop
12. Fabricated rebar storage yard
13. Carpentry shop
14. Cement warehouse
15. Batching plant and aggregates storage
16. Testing lab
17. Material storage lab
18. Water tank
19. Scaffolding storage
20. Canteen

Step 2 : Preparation of relationship chart.

The facilities are provided on temporary basis and related to their closeness with each other in this step.

The left hand side of the relationship diagram shows the list of identified temporary facilities and right hand of the chart shows the description of relation between the one facility to the another facility. Description of the relation between facilities is denoted by alphabets as A, B, C, D, E. The alphabet is also called as proximity weight or proximity value because there is fixed some relation of facility in respect to the remaining facilities depending on proximity relation.

A = Absolutely necessary that, the two facilities should be close (to be close).
B = Especially important that, the two facilities must be close.
C = Important that, the two facilities may be close.
D = Unimportant or no need to be close the two facilities.
E = Undesirable or need to be far the two facilities.

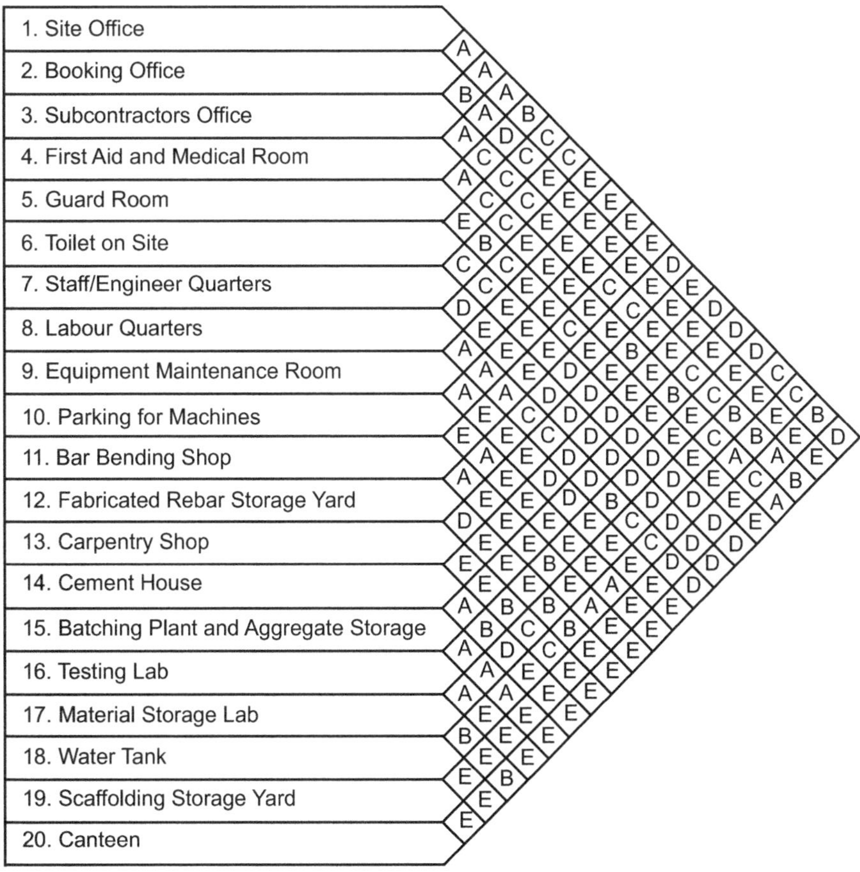

Relationship chart

Step 3 : Grouping of facilities.

The decision of fixing the facility location is depend on the entry location, exit and how will be the flow of facility occurs on site. The grouping of facilities is done according with their relationship.

For this case grouping is as :

Group 1 :
1. Site office.
2. Booking office.
3. Subcontractor office.
4. First aid and medical room.
5. Toilet on site.
6. Labour quarter.
7. Guard room.

Group 2 :
1. Toilet on site.
2. Carpentry shop.
3. Water tank.
4. Scaffolding storage.

Group 3 :
1. Cement warehouse.
2. Batching plant and aggregate storage.

Group 4 :
1. Toilet on site.
2. Staff and engineer quarter.

Group 5 :
1. Bar bending shop.
2. Fabricated rebar storage yard.

Group 6 :
1. Testing lab.
2. Material storage lab.

Group 7 :
1. Equipment maintenance room.
2. Parking for machines.

Group 8 :
1. Canteen.
2. Toilet on site.

Others :
1. Internal road.
2. Electric poles.

Step 4 : Space requirement of facilities.

Once the relationship between the temporary facilities provided and grouping of them are worked out, the space area requirement for individual facility is determined in accordance with available space.

For this case:

Sr. No.	Facility	Size A (in m)	Size B (in m)	Area (in m²)
1.	Site office	5	6	30
2.	Booking office	5	9	45
3.	Subcontractor's office	8	13	104
4.	First aid and medical room	5	6	30
5.	Guard room	6	3	18
6.	Toilet on site	28	4	112
7.	Engineer and staff quarters	12	20	240
8.	Labour quarters	12	55	660
9.	Equipment maintenance room	13	15	195
10.	Parking for machines	19	19	361
11.	Bar bending shop	7	25	175
12.	Fabricated rebar storage yard	4	12	48
13.	Carpentry shop	15	25	375
14.	Cement warehouse	10	30	300
15.	Batching plant and aggregates storage	16	22	352
16.	Testing lab	7	17	119
17.	Material storage lab	7	17	119
18.	Water tank	20	2	40
19.	Scaffolding storage	20	30	600
20.	Canteen	10	22	220

Finally the Job Layout diagram is as below :

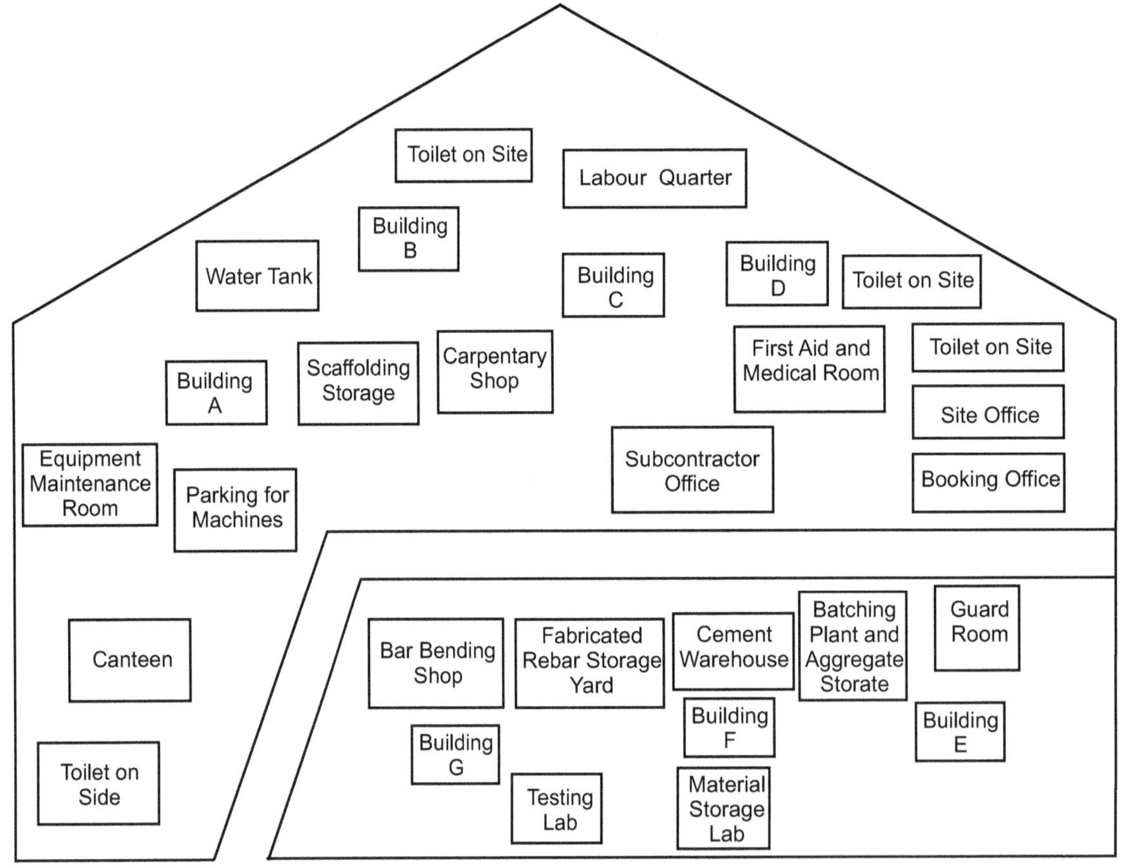

Experiment No. 11

Aim : Prepare charts/power point presentation on various safety devices used at construction site.

Hazards are of following categories :

1. Impact
2. Penetration
3. Compression
4. Chemical
5. Heat
6. Electrical shock
7. Electrical arc
8. Harmful dust
9. Light radiation
10. Falls

Places to Keep Safety on Construction Site :

1. Motion
2. Temperature
3. Chemical exposure
4. Light radiation
5. Elevation
6. Sharp objects
7. Rolling/pinching
8. Electrical hazards
9. Workplace layout
10. Worker Location

Protective Devices :

Following protective devices are need to be provided on construction site.

1. Head Protection :

Employees working in areas where there is a possible danger of head injury from impact, or from falling or flying objects, or from electrical shock and burns, shall be protected by helmets.

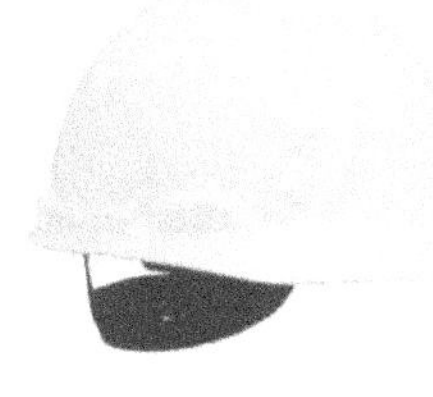

White: For Engineers, Managers, Supervisors and Foremen.

Blue: For Electricians, Carpenters and other Technical operators.

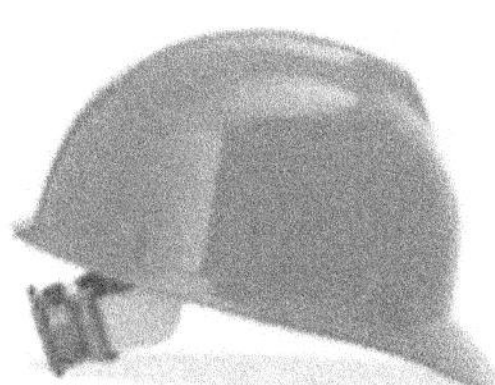

Red: For Fighters.

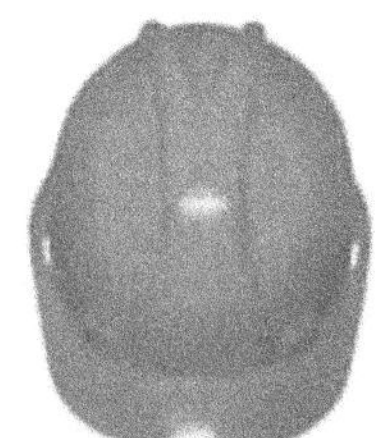

Green: For Safety Officers.

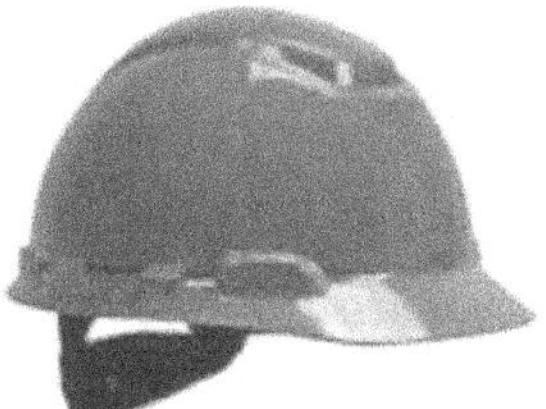

Gray: For Site Visitors.

Yellow: For Labourers and Earth Moving Operators.

Brown: For Welders and Workers with High Heat Application.

2. Hearing Protection :

Wherever it is not feasible to reduce the noise levels or duration of exposure, ear protection devices shall be provided and used. Ear protection devices inserted in the ear shall be fitted properly. Plain cotton is not an acceptable protective device.

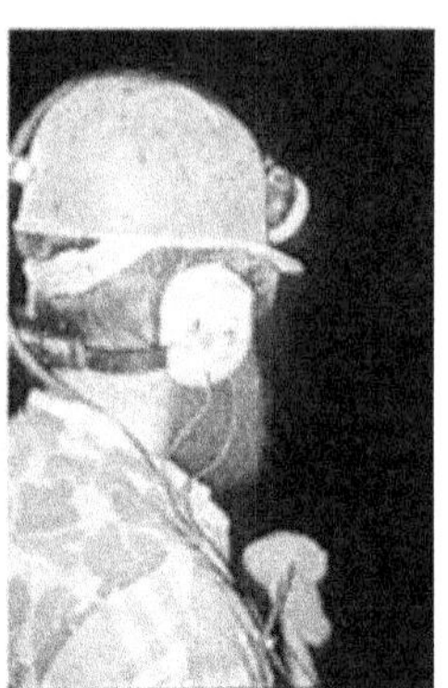

3. Eye and Face Protection :

Employees shall be provided with eye and face protection equipment when machines or operations present potential eye or face injury from physical, chemical, or radiation agents.

Employees whose vision requires the use of corrective lenses in spectacles, when required by this regulation to wear eye protection, shall be protected by goggles or spectacles.

- Spectacles whose protective lenses provide optical correction.
- Goggles that can be worn over corrective spectacles without disturbing the adjustment of the spectacles.
- Goggles that incorporate corrective lenses mounted behind the protective lenses.

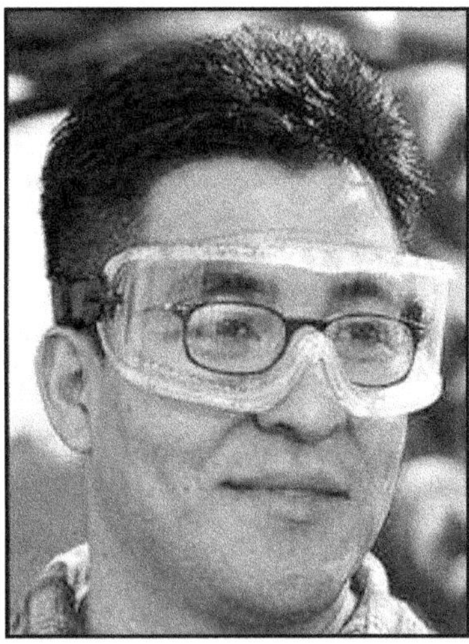

4. Foot Protection :

Safety-toe footwear for employees shall meet the requirements of site condition.

5. Respiratory Protection :

At the site condition prevailing to most unfavorable breathing, the employees working should be provided with proper respiratory protection.

6. Safety Belts, Lifelines, Lanyards :

Lifelines, safety belts, and lanyards shall be used only for employee safeguarding.

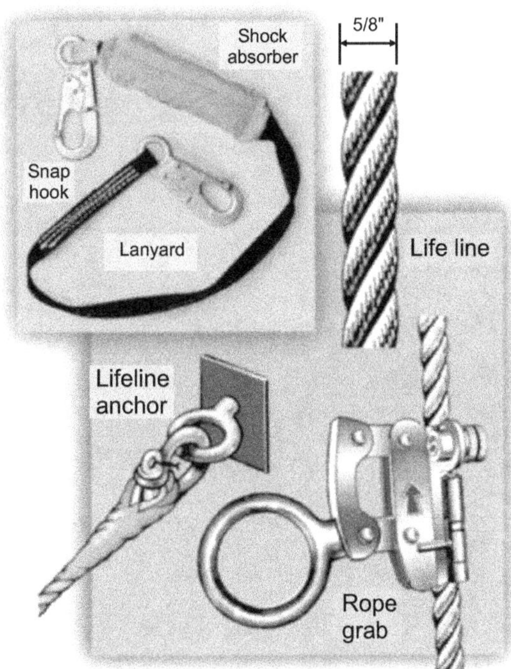

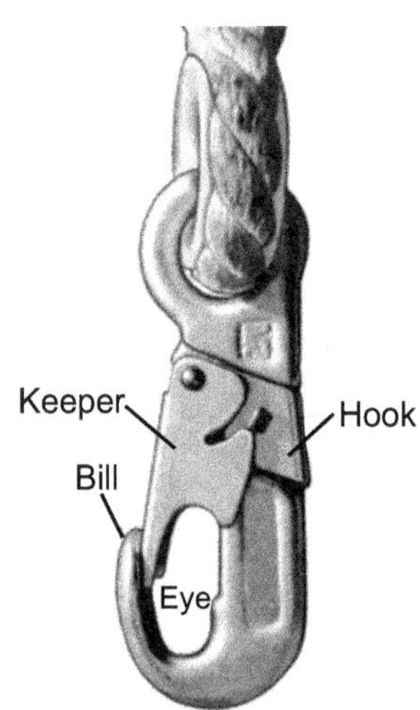

7. Safety Nets :

Safety nets shall be provided when workplaces are more than 25 feet above the ground, or water surface, or other surfaces where the use of ladders, scaffolds, catch platforms, temporary floors, safety lines, or safety belts is impractical.

□□□